Site Engineering
for Landscape Architects

Second Edition

Site Engineering for Landscape Architects

Second Edition

Steven Strom
Rutgers—The State University of New Jersey

Kurt Nathan PE
Conservation Engineering, P.A.

VNR VAN NOSTRAND REINHOLD
New York

Portions of Chapters 2, 3, 10, 11, 12 and 13 have been adapted from *Basic Site Engineering for Landscape Designers* by Kurt Nathan, © 1973, 1975 by MSS Information Corporation, transferred to Kurt.

Page 7, problem 1.3 is adapted with permission from V.C. Finch et al., *The Earth and Its Resources*, Third Edition, © 1959, by the McGraw-Hill Book Company, Inc., New York.

Page 47, quote from *The Penny Dales* by Arthur Raistrick is reprinted by permission of Methuen London Ltd. Publishers.

Page 106, the runoff coefficients for urban areas are taken with permission from Chapter 3, Storm Drainage, by Paul Theil Associates Ltd., in *Modern Sewer Design*, First Edition, © 1980 by the American Iron and Steel Institute.

Page 106, the runoff coefficients for rural and suburban areas are taken with permission from G.O. Schwab et al., *Elementary Soil and Water Engineering*, © 1971, by John Wiley & Sons, Inc., New York.

Page 108, Table 8.2 is taken with permission from H.G. Poertner, *Practices in Detention of Urban Storm Water Runoff*, American Public Works Association, Special Report No. 43, 1974.

Page 149, Figure 10.13 is redrawn with permission from Chapter 3, Storm Drainage, by Paul Theil Associates Ltd., in *Modern Sewer Design*, First Edition, © 1980 by the American Iron and Steel Institute.

Page 185, Table 12.1 is adapted with permission from K. Lynch, *Site Planning*, Second Edition, published by the MIT Press, © 1971 by the Massachusetts Institute of Technology.

Page 201, Figure E.3 and Page 202, Figure E.5c are published with the permission of Andropogon Associates.

Figures 9.1-9.9 and 9.13 and tables 9.1, 9.3-9.5, and 9.9a and b are from the USDA Soil Conservation Service *Urban Hydrology For Small Watersheds* (1986).

Copyright © 1993 by Van Nostrand Reinhold

Design and production: Editorial Services of New England, Inc.
Printed in the United States of America

All rights reserved. No part of this work covered by the copyright hereon may be reproduced or used in any form or by any means—graphic, electronic, or mechanical, including photocopying, recording, taping, or information storage and retrieval systems—without the written permission of the publisher.

98 99 COUWF 10 9 8 7

Library of Congress Cataloging-in-Publication Data
Strom, Steven.
    Site engineering for landscape architects / Steven Strom, Kurt Nathan.—2nd ed.
      p.   cm.
    Includes bibliographical references(p.) and index.
    ISBN 0-442-00224-6
    1. Building sites. 2. Landscape architecture. I. Nathan, Kurt.
  II. Title.
TH375.S77 1992
690'.11—dc20   91-45494
                                        CIP

To
Beth, Matthew, Peter and Emily Strom
and
Barbara and B. David Nathan

Contents

Preface xi
Acknowledgments xi

1 Contours and Form 1

Definition 1
Construction of a Section 3
Contour Signatures and Landform 3
Characteristics of Contour Lines 6
Exercises 7

2 Interpolation and Slope 9

Topographic Data 9
Interpolation 9
Calculation of Slope 13
Slopes Expressed as Ratios and
  Degrees 15
Exercises 16

3 Slope Formula Applications 17

Slope Analysis 17
Slopes for Surface Drainage 18
Terrace Grading 20
Path Layout with a Maximum Gradient 24
Grading of Roads 25
Visualization of Topography from Contour
  Lines 29
Exercises 30

4 Grading Constraints 35

Environmental Constraints 35
Functional Constraints 38
Summary of Critical Constraints 43
Exercises 43

5 Grading Design and Process 45

Grading Design 45
Grade Change Devices 48
Slopes 52
Grading Process 53
Grading Plan Graphics 60
Exercises 61

6 Grading and Landform Design: Case Studies 67

Introduction 67
Earthworks Park 67
Gasworks Park 70
Olympic Park 73
Westpark 79
Exercises 83

7 Storm Water Management 85

Storm Runoff 85
Hydrologic Cycle 85
Nature of the Problem 85

Management Philosophy   87
Principles and Techniques   88
Control of Erosion and Sedimentation   96
Summary   99
Exercises   99

8 Determination of Rates and Volumes of Storm Runoff: The Rational and Modified Rational Methods   101

Introduction   101
Rational Method   101
Modified Rational Method   108
Volumes of Runoff, Storage, and Release   112
Required Storage for Detention or Retention Ponds by the Modified Rational Method   113
Summary   114
Exercises   115

9 Soil Conservation Service Methods of Estimating Runoff Rates, Volumes, and Required Detention Storage   117

Introduction   117
Rainfall   117
Procedures of TR55   117
Volume for Detention Storage   128
Summary   132
Exercises   132

10 Design and Sizing of Storm Water Management Systems   133

Management Systems   133
Design and Layout of Drainage Systems   137
Applications   138
Subsurface Drainage   152
Use of Computers   155
Summary   155
Exercises   155

11 Earthwork   159

Definitions   159
Construction Sequence for Grading   160
Grading Operations   161
Computation of Cut and Fill Volumes   162
Exercises   171

12 Horizontal Road Alignment   175

Types of Horizontal Curves   175
Circular Curve Elements   176
Circular Curve Formulas   176
Degree of Curve   178
Stationing   179
Horizontal Sight Distance   182
Construction Drawing Graphics   183
Horizontal Alignment Procedures   183
Superelevation   185
Exercises   186

13 Vertical Road Alignment   189

Vertical Curve Formula   190
Unequal Tangent Curves   190
Equal Tangent Curves   191
Calculation of Locations of High and Low Points   192
Construction Drawing Graphics   193
Vertical Sight Distances   194
Road Alignment Procedure   194
Exercises   196

Epilogue   199

Glossary   205

Bibliography   209

Index   211

Preface

The shaping of the earth's surface is one of the primary functions of site planners and landscape architects. This shaping must display not only sound understanding of aesthetic and design principles but also ecological sensitivity and technical competency. The last point, the technical ability to transform design ideas into physical reality, is the focus of this text. Specifically the book emphasizes principles and techniques of basic site engineering for grading, drainage, earthwork, and road alignment. The authors strongly believe that, in most cases, it is difficult to separate the design procedure from the construction and implementation process and that technical competency ultimately will lead to a better finished product. In this regard the authors also feel that collaborative efforts between the landscape architecture and engineering professions will result in the most appropriate solutions to design and environmental problems.

Although appropriate as a reference for practitioners, the book has been developed primarily as a teaching text. The material, numerous examples, and problems, which are based on the authors' experience, have been organized to provide students with a progressive understanding of the subject matter. First, landform and contour lines are discussed descriptively. This is followed by explanations of interpolation and slope formulas and examples of their applications. Chapters 4 and 5 are concerned with environmental and functional constraints and design opportunities which guide site engineering decisions, and a procedure for developing grading solutions. The integrated relationship between site design and site engineering is also emphasized in these chapters. Case studies which demonstrate the successful application of grading and landform design as important site planning techniques are presented in Chapter 6. Storm water management and the design and sizing of management systems are discussed in Chapters 7, 8, 9, and 10, with particular emphasis on the Rational, Modified Rational, and Soil Conservation Service methodologies. Earthwork terminology, construction sequencing, and computation of earthwork volumes are presented in Chapter 11. The procedures for designing horizontal and vertical road alignments are presented in the last two chapters. In summary, the authors believe that the book provides a strong foundation in the essential aspects of site engineering.

ACKNOWLEDGMENTS

Many sources have contributed to the authors' interest and growth in the field of site development and landscape construction. The classic text by Parker and MacGuire and the work of David Young and Donald Leslie deserve special mention among them. Publications of the USDA Soil Conservation Service, particularly in the areas of storm water management and swale design, have been invaluable.

Thanks are due to the Department of Landscape Architecture, Cook College, Rutgers—The State University of New Jersey, for the use of facilities and equipment; to Margarita Bermudez for her assistance in typing the manuscript for this edition; to Elizabeth Grady and Karneel Thomas for their help with the illustrations of the first edition; and to Kathleen John-Alder for the plan drawings in Chapter 6 of the second edition. We are grateful to Günther Grzimek, Richard Haag, Peter Kluska, and Nancy Leahy for the information which they provided for the case studies in Chapter 6, and to Andropogon Associates for the information on the Morris Arboretum.

Robert Moore supplied the photograph for Fig. 10.7b.

Leland Anderson and Lee Holt of the Soil Conservation Service (SCS) reviewed several portions of the manuscript for the first edition. David Lamm, also of the SCS, reviewed the entire text of the first edition and Chapters 7, 8, 9, and 10 of this edition. Vincent Abbatiello and Guy Metler performed the photographic processing for the first edition. Their assistance is gratefully acknowledged.

Finally, the authors would like to express their thanks to all who have made helpful suggestions for improving the text, especially Marvin Adleman and John Roberts.

1
Contours and Form

DEFINITION

A clear understanding of what a contour represents is fundamental to the grading process. Technically defined, a *contour* is an imaginary line that connects all points of equal elevation above or below a fixed reference plane or datum. This datum may be mean sea level or a locally established bench mark. A *contour line* is the graphic representation of a contour on a plan or map. Within this text, however, the terms *contour* and *contour line* will be used synonymously.

A difficulty with understanding contours arises from the fact that they are imaginary and therefore cannot be easily visualized in the landscape. The shoreline of a pond or lake is the best example of a naturally occurring contour and illustrates the concept of a closed contour. A *closed contour* is one that reconnects with itself. All contours eventually close on themselves, although this may not occur within the boundaries of a particular map or plan.

A single, closed contour may describe a horizontal plane or level surface, again illustrated by a pond or lake. However, more than one contour is required to describe a three-dimensional surface. Rows of seating in an athletic stadium or amphitheater (Fig. 1.1a) provide an excellent way to visualize a series of contours which define a bowl-shaped form. It is important to emphasize here that contour drawings are two-dimensional representations of three-dimensional forms. A basic skill that landscape architects and site designers must develop is the ability to analyze, interpret, and visualize landforms from contour maps and plans, commonly referred to as *topographic maps*. Designers must understand not only existing contours and landforms but the implications of changes, both aesthetic and ecological, which result from *altering* contours. The series of illustrations in Figs. 1.2 and 1.3 demonstrate how contours define form and how a form may be altered by changing contours. The contour plan of the

FIGURE 1.1. Visualizing Contour Lines. a. The stepped levels formed by stadium benches demonstrate the concept of contour lines. b. Small terraces created for spectators at the site of the 1972 Olympic kayak run provide an excellent example of contour lines as they appear on the ground.

2   Contours and Form

 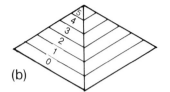 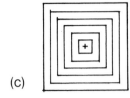

FIGURE 1.2. Relationship of Contour Lines to Three-Dimensional Form. a. Isometric drawing of pyramidal form. b. Contour lines illustrated on the isometric drawing. c. Contour plan of pyramid (concentric squares).

pyramid results in a series of concentric squares. By changing the squares to circles, the form is redefined from a pyramid to a cone. Figure 1.3 illustrates this transformation, starting with the contour plan.

Another aspect of contours and form is illustrated by Fig. 1.3b and 1.3c and by Fig. 1.4. A *gradual* rather than abrupt change is assumed to occur between adjacent contours. In Fig. 1.4, a section (see the definition in the following section) has been taken through the center of the cone. (Note that a section taken through the center of the pyramid results in the same two-dimensional form.) The steplike form that results from stacking the successive planes is indicated by the dashed line, and the smoothing effect that results from assuming a gradual transition is indicated by the shaded triangles. It is this smoothing effect that gives the cone and pyramid their true form. Again stadium seating provides a good example of the steplike character created by adjacent contours where a smooth transition has not been taken into consideration.

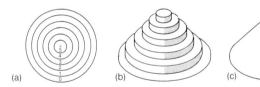

FIGURE 1.3. Alteration of Form by Changing Contour Lines. a. Square contour lines of pyramid altered to concentric circular contour lines. b. Horizontal planes of circular contours stacked in layer-cake-like manner. c. Isometric of resultant conical form.

These examples are overly simplistic in their approach, since they deal with basic geometric forms and straightforward alterations. However, the landscape consists of numerous geometric shapes occurring in complex combinations. The ability to dissect landforms into their various component shapes and to understand the relationship of the shapes to each other will make the task of analyzing, interpreting, and visualizing the landscape easier.

The difference in elevation between adjacent contour lines as illustrated by the steps in Fig. 1.4 is defined as the *contour interval*. In order to interpret a topographic map properly, scale, direction of slope, and contour interval must be known. The most common intervals are 1, 2, 5, 10, and multiples of 10 ft. Selection of a contour interval is based on the roughness of the terrain and the purpose for which the topographic plan is to be used. It is obvious that as the map scale decreases (for example, changing from 1 in. = 20 ft to 1 in. = 100 ft for the same area) or the contour interval increases, the amount of detail, and therefore the degree of accuracy, decreases (Fig. 1.5).

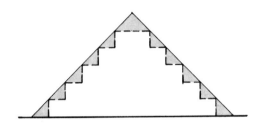

FIGURE 1.4. Surface Smoothing between Adjacent Contour Lines.

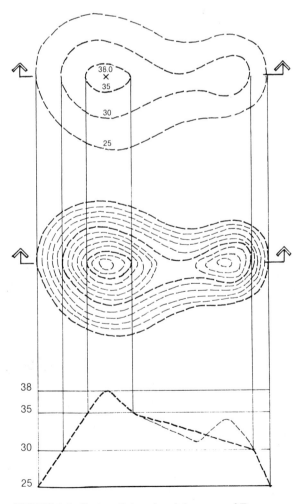

FIGURE 1.5. Contour Interval and Accuracy of Form.

CONSTRUCTION OF A SECTION

Analyzing topography and landform can be accomplished by constructing a section. A *section* is a drawing made on a plane, which vertically cuts through the earth, an object like a building, or both. The ground line delineates the interface between earth and space and illustrates the relief of the topography. To draw a section, follow the procedure outlined in Fig. 1.6.

In Fig. 1.6 the highest elevation of the landform occurs between the 13-ft and 14-ft elevations. Therefore, a peak, or high point, must occur between the two intersections along the 13-ft elevation line. A similar condition, and how it may be misinterpreted as a result of degree of accuracy, is illustrated in Fig. 1.5.

CONTOUR SIGNATURES AND LANDFORM

It becomes apparent in analyzing landform that certain geomorphic features are described by distinct contour configurations. These configurations may be referred to as *contour signatures*. Typical contour signatures are identified on the contour maps (portions of United States Geological Survey [USGS] quadrangles) in Figs. 1.7 and 1.8.

Ridge and Valley

A *ridge* is simply a raised elongated landform. At the narrow end of the form the contours point in the downhill direction. Typically, the contours along the sides of the ridge will be relatively parallel and there will be a high point or several high points along the ridge.

A *valley* is an elongated depression that forms the space between two ridges. Essentially valleys and ridges are interconnected, since the ridge side slopes create the valley walls. A valley is represented by contours that point uphill.

Contour patterns are similar for ridges and valleys; therefore, it is important to note the direction of slope. In each case the contours reverse direction to create a U or V shape. The V shape is more likely to be associated with a valley, since the point at which the contour changes direction is the low point. Water collects along the intersection of the sloping sides and flows downhill, forming a natural drainage channel at the bottom.

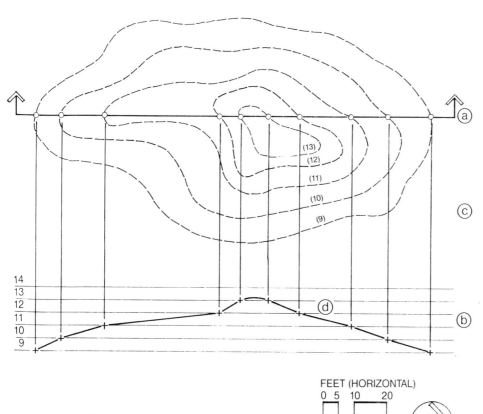

FIGURE 1.6. Drawing a Section.
a. Indicate cutting plane.
b. Draw parallel lines according to contour interval and proposed vertical scale.
c. Project perpendicular lines from the intersection of the contour line with the cutting plane to the corresponding parallel line.
d. Connect the points to complete the section and delineate the ground line.

4 Contours and Form

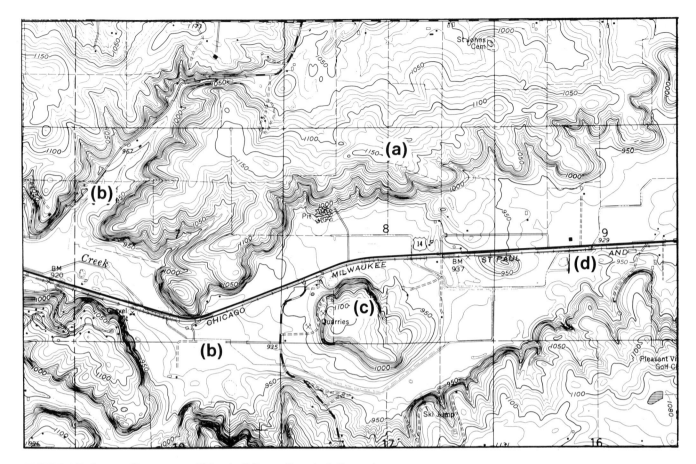

FIGURE 1.7. Contour Signatures. a. Ridge. b. Valley. c. Summit. d. Depression.

Summit and Depression

A *summit* is a landform such as a knoll, hill, or mountain, which contains the highest point relative to the surrounding terrain. The contours form concentric, closed figures with the *highest* contour at the center. Since the land slopes away in all directions, summits tend to drain well.

A *depression* is a landform that contains the lowest point relative to the surrounding terrain. Again the contours form concentric, closed figures, but now the *lowest* contour is at the center. To prevent confusion between summits and depressions it is important to know the direction of elevation change. Graphically, the lowest contour is often distinguished by the use of hachures. Since depressions collect water, they typically form lakes, ponds, and wetlands.

Concave and Convex Slopes

A distinctive characteristic of *concave* slopes is that the contour lines are spaced at *increasing* distances in the downhill direction. This means that the slope is steeper at the higher elevations and becomes progressively more flat at the lower elevations.

A *convex* slope is the reverse of a concave slope. In other words, the contour lines are spaced at *decreasing* distances in the downhill direction. The slope is flatter at the higher elevations and becomes progressively more steep at the lower elevations.

Uniform Slope

Along a *uniform slope,* contour lines are spaced at *equal* distances. Thus, the change in elevation occurs at a constant rate. Uniform slopes are more typical in constructed than in natural environments.

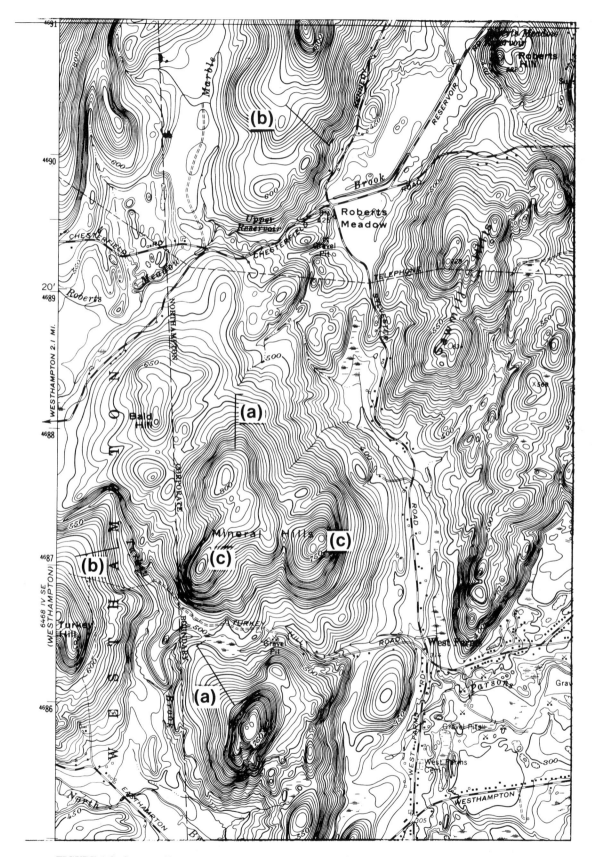

FIGURE 1.8. Contour Signatures. a. Concave slope. b. Convex slope. c. Summit.

6 Contours and Form

CHARACTERISTICS OF CONTOUR LINES

The following points summarize the essential characteristics associated with contour lines. Since many of the concepts and principles discussed in subsequent chapters relate to these characteristics, a thorough understanding must be achieved before proceeding.

1. By definition, all points on the same contour line are at the same elevation.
2. Every contour line is a continuous line, which forms a closed figure, either within or beyond the limits of the map or drawing (Fig. 1.9).
3. Two or more contour lines are required to indicate three-dimensional form and direction of slope (Fig. 1.10).
4. The steepest slope is perpendicular to the contour lines. This is a result of having the greatest vertical change in the shortest horizontal distance. (Fig. 1.10)
5. Consistent with the preceding point, water flows perpendicular to contour lines.
6. For the same scale and contour interval, the steepness of slope increases as the map distance between contour lines decreases.
7. Equally spaced contour lines indicate a constant, or uniform, slope.
8. Contour lines never cross except where there is an overhanging cliff, natural bridge, or other similar phenomena.
9. In the natural landscape, contour lines never divide or split. However, this is not necessarily true at the interface between the natural and built landscape, as illustrated in Fig. 1.11.

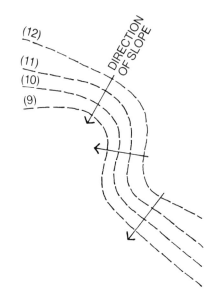

FIGURE 1.10. Direction of Slope. The steepest slope is perpendicular to the contour lines. Consequently, surface water flows perpendicular to contour lines.

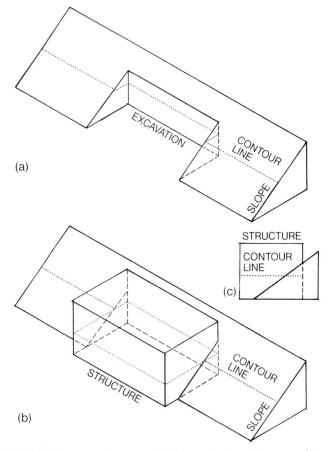

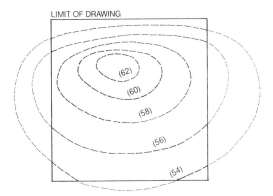

FIGURE 1.9. Closed Contours. Contours are continuous lines creating closed figures. However, closure may not always occur within the limits of a drawing or map.

FIGURE 1.11. Technically, contour lines never divide or split where they are used to represent the surface of the earth. However, at structures contour lines may also be drawn across the face of the constructed object, thus providing a split appearance. a. The contour line follows along the face of an excavation made in a slope. b. The contour line follows along the face of the excavation as well as along the face of the structure placed in the excavated area. c. End section illustrating the relationship between the slope and the structure.

EXERCISES

1.1 The intent of this two-part problem is to develop your ability to visualize landform from contours. (a) Draw a contour plan of the landform shown in Fig. 1.12. Use a minimum of eight contour lines to depict the form. (b) Draw an oblique aerial perspective of the landform represented by the contour plan shown in Fig. 1.13.

1.2 Exercise 1.1 required the visualization of landforms and contour lines using two-dimensional graphics. An easier, but more time-consuming, method for interpreting contours is through the use of three-dimensional models.

Construct two models: the first of the contour plan in Exercise 1.1b and the second of the more architectural landform illustrated in Fig. 1.14. Once constructed, these models may be used to analyze various contour line relationships: relative steepness, concave and convex slopes, and so on.

1.3 As an alternative to Exercise 1.2 mold an oval mound of wax 6½ in. high, steeply sloping at one end and gently sloping at the other in an open tank. Fill the tank with 6 in. of water, thus leaving ½ in. of the mound protruding. With a sharp point outline the position of the edge of the water on the wax. Next lower the level of the water by 1-in. stages, outlining the position of the water at each stage until all water has been removed. The marks made will now appear as contour lines on the wax mound, the lowest being 1 in. from the bottom of the tank, the next 2 in., and so on. If the mound is viewed from directly above, the arrangement of the lines is exactly like that of a contour map. Notice that where the slope of the mound is steep, the contour lines are close together and that as the slope becomes more gentle, they are more widely spaced.

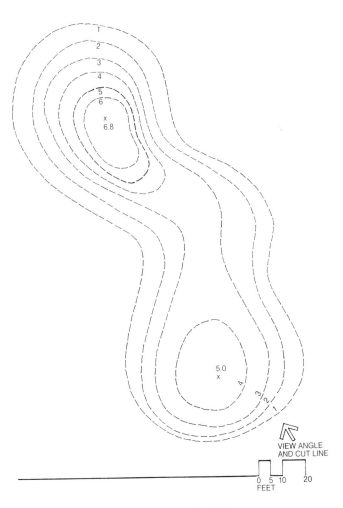

FIGURE 1.13. Contour Plan for Exercise 1.1b.

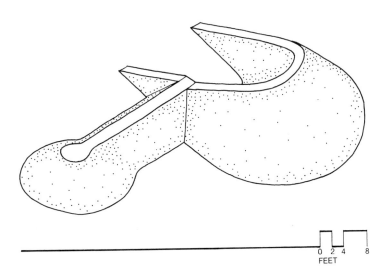

FIGURE 1.12. Axonometric of Landform for Exercise 1.1a.

1.4 The first two exercises address the issue of understanding the three-dimensional forms created by contour lines using graphic and model techniques. However, these methods are still somewhat abstract, since they are not related directly to the landscape. There are two techniques that may be used to place contour lines and form in a realistic context: (a) "drawing" contour lines directly on the ground by the use of lime, etching lines in snow (if the weather is appropriate), or by the use of string or surveyor's flagging and, (b) drawing contour lines on a map from an on-site visual analysis. Select a small area with a variety of topographic conditions and attempt one or both of these techniques. As a clue to laying out contour lines, there are numerous features in the landscape that can help to determine relative differences in slope and elevation. These include stairs, brick courses on buildings, door heights, vegetation, and people. Keying on these features will make this task easier.

1.5 Construct a section of the landform in Exercise 1.1b along the cut line indicated. Use 1 in. = 10 ft for the horizontal scale and 1 in. = 5 ft for the vertical scale.

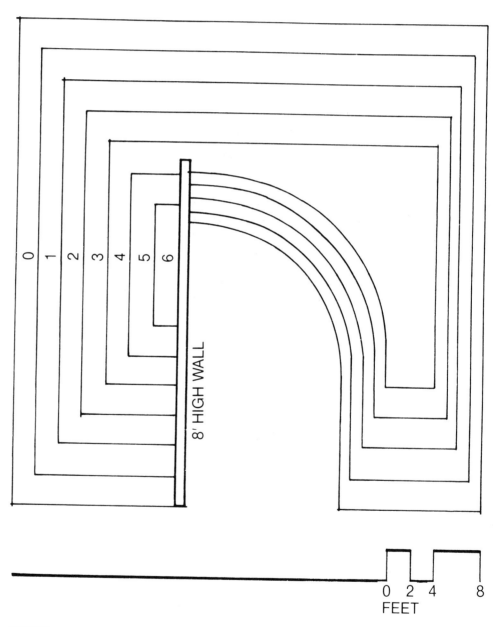

FIGURE 1.14. Contour Plan for Exercise 1.2.

2
Interpolation and Slope

TOPOGRAPHIC DATA

The previous chapter discussed contour lines from a descriptive viewpoint: that is, explaining what contour lines portray. This chapter introduces the basic mathematical equations associated with plotting and manipulating contour lines.

In order to make informed design decisions as well as to execute construction drawings accurately, landscape architects require topographic data for all site development projects. This information is usually provided by a licensed land surveyor in the form of a topographic map or plan. However, there are occasions when a landscape architect may be required to gather topographic data, or the data furnished by the surveyor are merely spot elevations, usually on a rectangular grid. This section discusses the latter two points.

Typically, topographic data are collected by laying out a grid pattern over the site to be surveyed. The size of the grid selected depends on the extent of the area, the degree of topographic variation, and the purpose for which the survey will be used. Generally 20-, 25-, 50-, or 100-ft squares are used. For large rectangular areas, laying out the grid is a relatively simple task. Two rows of stakes may be placed along each of two sides of the rectangle, as illustrated in Fig. 2.1, thus allowing the rod person to locate the remaining grid intersection points by aligning the leveling rod with two pairs of stakes. Leveling proceeds as in differential leveling except that a number of foresights can be taken for each instrument setup. If the elevation of the area does not vary more than 10 or 12 ft, it may be possible to obtain all necessary data with one instrument setup. For referencing purposes, the grid intersections are labeled with letter and number coordinates.

For sites that are more complex, in configuration, topographic variety, or both, the same basic principles may be applied with modifications, such as additional setups, more foresights, or a grid geometry compatible with the shape of the site. Further foresights may be necessary where a high or low point exists between grid intersections since, as discussed in the next section, the assumption is that there is a constant or uniform change in elevation from one grid corner to the next.

Once the elevations for each of the grid corners have been determined, they are plotted on a plan at a desired or specified scale. The next step in the process is to draw the contour lines (Fig. 2.6) since, as discussed in Chapter 1, this allows for easier visualization and understanding of the three-dimensional form of the landscape. Before drawing the contour lines, however, it is necessary to introduce the concept of interpolation.

INTERPOLATION

Interpolation, by definition, is the process of computing intermediate values between two related known values. For the purpose of topographic surveys, interpolation is the process of locating whole number elevations (assuming that contour lines with a 1-ft vertical interval are desired) between the elevations of the grid intersections. Interpolation may be performed by constructing a simple proportional equation

$$\frac{d}{D} = \frac{e}{E} \qquad (2.1)$$

where d = distance from one grid intersection to an intermediate point, ft

  D = total distance between grid intersections, ft

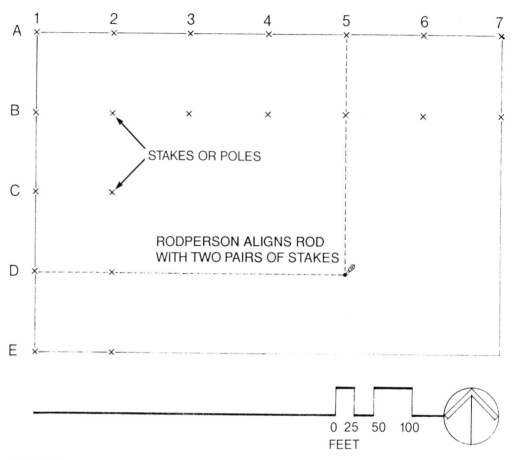

FIGURE 2.1. Plan of Stake Layout for Collecting Topographic Data. As discussed in Chapter 11, the grid of spot elevations is particularly helpful when using the borrow pit method of determining cut and fill volumes. However, irregular grids or selected points may also be used to collect topographic data.

e = elevation change between the same grid intersection and the intermediate point, ft
E = total elevation change between grid intersections, ft

Interpolation Between Spot Elevations

The sample grid cell in Fig 2.2. and accompanying calculations illustrate the interpolation process. The change in elevation along side a of the cell is from 97.3 to 95.3, for a total difference of 2.0 ft. Between these two points are the 97.0 and 96.0 elevations, which must be located in order to draw the whole number contours for the cell. The difference in elevation from 97.3 to 97.0 is 0.3 ft. Together with the knowledge that the cell measures 100 ft on each side, three parts of the equation are known. Substituting into the proportional equation, the unknown value, which is the distance d from the 97.3 spot elevation to the 97.0 spot elevation, can be calculated:

$$\frac{d}{100} = \frac{0.3}{2.0}$$

$$d \times 2.0 = 0.3 \times 100$$

$$d = \frac{0.3 \times 100}{2.0} = 15.0 \text{ ft}$$

Measuring 15.0 ft from 97.3 with the appropriate scale will locate the 97.0 elevation along side a.

The difference between 97.3 and 96.0 is 1.3 ft. Substituting into the equation, the distance from the 97.3 spot elevation to the 96.0 spot elevation is 65.0 ft.

$$\frac{d}{100} = \frac{1.3}{2.0}$$

$$d \times 2.0 = 1.3 \times 100$$

$$d = \frac{1.3 \times 100}{2.0} = 65.0 \text{ ft}$$

Note that the distance also could have been determined from the 95.3 spot elevation. The difference in elevation from 95.3 to 96.0 is 0.7 ft, which computes to a distance of 35.0 ft, but now measured from the 95.3 intersection. As a check, the two distances calculated for the 96.0 spot elevation when added together must total the dimension of the side of the grid cell (65.0 + 35.0 = 100.0 ft).

The calculations for the remaining three sides are summarized next and the completed cell is shown in Fig. 2.3.

Side b:

$98.5 - 97.3 = 1.2$ ft (total elevation change, E)

$98.5 - 98.0 = 0.5$ ft (elevation change, e)

$$\frac{d}{100} = \frac{0.5}{1.2}$$

$$d = \frac{0.5}{1.2} \times 100 = 41.67 \text{ ft from the 98.5 intersection.}$$

Side c:

$98.5 - 96.9 = 1.6$ ft (E)

$98.5 - 98.0 = 0.5$ ft (e)

$$\frac{d}{100} = \frac{0.5}{1.6}$$

$$d = \frac{0.5}{1.6} \times 100 = 31.25 \text{ ft from the 98.5 intersection}$$

$98.5 - 97.0 = 1.5$ ft (e)

$$\frac{d}{100} = \frac{1.5}{1.6}$$

$$d = \frac{1.5}{1.6} \times 100 = 93.75 \text{ ft from the 98.5 intersection}$$

Side d:

$96.9 - 95.3 = 1.6$ ft (E)

$96.9 - 96.0 = 0.9$ ft (e)

$$\frac{d}{100} = \frac{0.9}{1.6}$$

$$d = \frac{0.9}{1.6} \times 100 = 56.25 \text{ ft from the 96.9 intersection}$$

Rather than calculating each elevation, interpolation may be done graphically, as illustrated in Fig. 2.4. Along side *b* of the grid cell there is a difference in elevation of 1.2 ft. By dividing the side into 12 equal spaces (each space representing 0.1 ft of elevation change), counting either 7 spaces from the 97.3 spot elevation or 5 spaces from 98.5, the 98.0 spot elevation is located. The side of any grid cell may be divided into the desired number of *equal* spaces by placing a scale *at any angle* with the appropriate number of divisions (usually the same as the desired number of equal spaces) between perpendicular lines extended from the two end points of the side of the grid cell. From the appropriate point on the scale a line perpendicular to the side of the grid cell is extended back to the cell, thus locating the desired spot elevation. This is a common technique for graphic proportioning and may be applied to the other three sides of the grid cell.

Figure 2.5 represents the completed grid for the layout illustrated in Fig. 2.1. The process of interpolation may be applied to each of the grid cells to locate all the whole number spot elevations. (The sample grid cell is taken from the northwest corner of the grid.) Connecting the points of equal elevation may be left until the locations of all the whole number elevations have been determined for the entire grid, since contours are usually smooth curves rather than a series of straight line segments. The completed contour plan is shown in Fig. 2.6. Construction lines, including the grid, commonly are not shown on the completed plan. The closed 95-ft contour in Fig. 2.6 is a depression, which is depicted by the use of hachures, as illustrated.

With practice, contour lines can be drawn quite rapidly on a grid, with much of the interpolation done mentally and visually. It should be noted here that computers in conjunction with digitizers and plotters may be used to generate contour plans, thus eliminating the need for tedious calculations or drafting time.

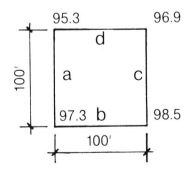

FIGURE 2.2. Sample Grid Cell.

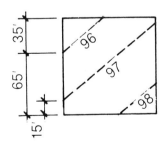

FIGURE 2.3. Sample Grid Cell with Contour Lines Located.

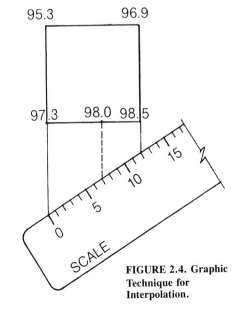

FIGURE 2.4. Graphic Technique for Interpolation.

FIGURE 2.5. Grid of Spot Elevations.

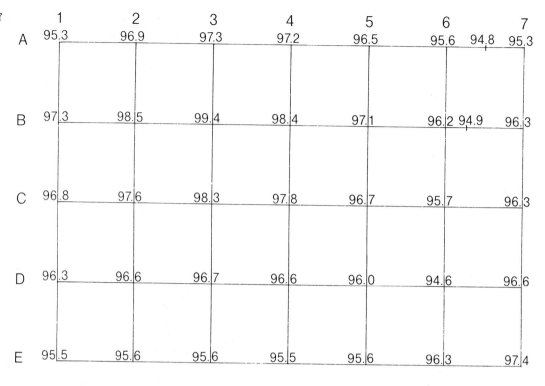

FIGURE 2.6. Contour Plan Interpolated from Fig. 2.5.

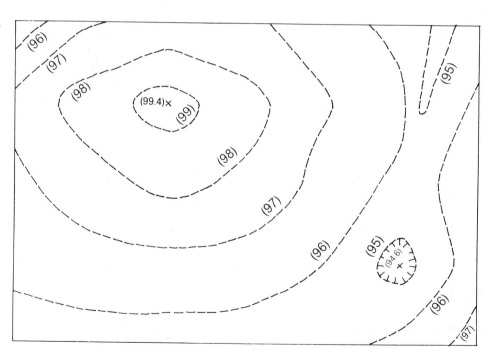

Interpolation Between Contour Lines

Interpolation may also be used to determine the elevation of points between contour lines. The information needed to compute these points includes the contour interval, the total distance between the contour lines, and the distance from one contour line to the point in question. With this information, the difference in elevation from the contour line to the point can be calculated by the following equation:

$$\frac{\text{distance from point to contour line}}{\text{total distance between contour lines}} \times \text{contour interval} = \text{elevation difference}$$

Example 2.1

Determine the spot elevation for point A in Fig. 2.7. The contour interval is 1 ft, and the distances are as indicated on the drawing.

Solution. The difference in elevation between point A and the 67-ft contour line is calculated by substituting into the equation

$$\frac{4.0 \text{ ft}}{10.0 \text{ ft}} \times 1.0 \text{ ft} = 0.4 \text{ ft}$$

The spot elevation at point A is then determined by adding the elevation difference to the elevation of the contour line used as the point of reference.

$67.0 + 0.4 = 67.4$ ft (elevation point A)

Example 2.2

Determine the spot elevation for point B in Fig. 2.8. The contour interval is 5 ft, and the distances are as indicated on the drawing.

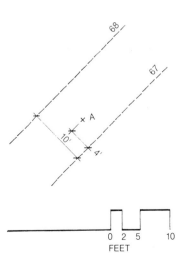

FIGURE 2.7. Plan for Example 2.1.

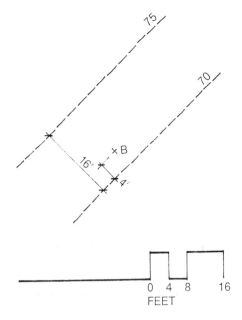

FIGURE 2.8. Plan for Example 2.2.

Solution. Using the same procedure as in Example 2.1, the difference in elevation is calculated as follows:

$$\frac{4.0 \text{ ft}}{16.0 \text{ ft}} \times 5.0 \text{ ft} = 1.25 \text{ ft}$$

By adding the difference in elevation to the 70-ft contour line, the spot elevation at point B is determined.

$70.0 + 1.25 = 71.25$ ft (elevation point B)

Finally it must be emphasized that the interpolation process is valid only if there is a constant slope between two points, whether those points are contours or spot elevations.

CALCULATION OF SLOPE

Most often, changes in grade are described or discussed in terms of percentage of slope. Describing slope in this manner provides a common basis of understanding for the professionals associated with manipulating and changing the earth's surface, including landscape architects, engineers, and architects. *Slope,* expressed as percentage, is the number of feet of rise or fall in 100 ft of horizontal distance. In addition to being expressed as a percentage, slope may also be described by decimal number. For example, a 12% slope may also be called a 0.12 slope, but *not* a 0.12% slope. The terms *grade* and *gradient* are commonly used synonymously with *slope*.

A generalized definition of slope then is the number of feet of fall or rise in a horizontal distance or $S = DE/L$, where S is the slope and DE is the difference in elevation between the end points of a line of which the horizontal or map distance is L (Fig. 2.9). To express S as a percentage, multiply the value by 100.

14 Interpolation and Slope

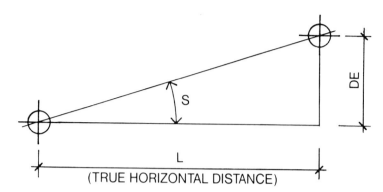

$$S = \frac{DE}{L}$$

DE = DIFFERENCE IN ELEVATION IN FT
L = HORIZONTAL DISTANCE IN FT
S = GRADIENT, EXPRESSED AS PERCENTAGE

FIGURE 2.9. Diagram of Slope Formula.

One problem that commonly arises is realizing that L is measured horizontally rather than along the slope. To reinforce this point, it should be remembered from surveying that all map distances are measured horizontally and not parallel to the surface of sloping ground.

With the slope formula, three basic computations may be accomplished:

1. Knowing the elevations at two points and the distance between the points, *slope S* can be calculated.
2. Knowing the difference in elevation between two points and the percentage of slope, the *horizontal distance L* can be calculated.
3. Knowing the percentage of slope and the horizontal distance, the *difference in elevation DE* can be calculated.

The following sample problems illustrate the three basic applications of the slope formula.

Example 2.3

Two spot elevations are located 120 ft apart (measured horizontally). One spot elevation is 44.37 ft; the other is 47.81 ft. (Fig. 2.10). Calculate the percentage of slope S.

Solution. First determine the difference in elevation DE between the spot elevations.

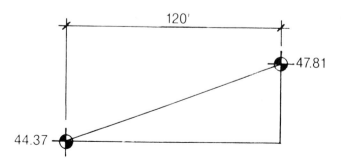

FIGURE 2.10. Section for Example 2.3.

$DE = 47.81 - 44.37 = 3.44$ ft

Then substitute the known values into the slope formula:

$$S = \frac{DE}{L} \qquad (2.2)$$

$$= \frac{3.44}{120.0} = 0.0287 \text{ ft/ft} = 2.87\%$$

Example 2.4

Determine the location of the whole number spot elevations (i.e., 45.0, 46.0, 47.0) in the previous problem.

Solution. Since the horizontal distance L is the unknown, the slope formula may be rearranged as follows:

$$L = \frac{DE}{S} \qquad (2.3)$$

With S previously determined in Example 2.3, the next step is to calculate the difference in elevation between the known and desired spot elevations and substitute the values into the preceding equation. (Fig. 2.11).

$DE = 47.81 - 47.00 = 0.81$ ft

$$L = \frac{0.81}{0.0287} = 28.26 \text{ ft}$$

$DE = 47.00 - 46.00 = 1.00$ ft

$ = 46.00 - 45.00 = 1.00$ ft

$$L = \frac{1.00}{0.0287} = 34.88 \text{ ft}$$

$DE = 45.00 - 44.37 = 0.63$ ft

$$L = \frac{0.63}{0.0287} = 21.98 \text{ ft}$$

As a check, the sum of the partial distances should equal 120 ft. (Note that in computing these distances, the value used for S was not rounded.)

$28.26 + 34.88 + 34.88 + 21.98 = 120.00$ ft

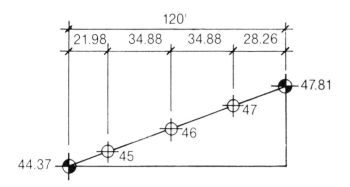

FIGURE 2.11. Section for Example 2.4.

Example 2.5

Determine the spot elevation of a point 40 ft uphill from elevation 44.37. Use the slope calculated in Example 2.3.

Solution. Since the difference in elevation *DE* is the unknown, the equation may be rearranged as follows:

$$DE = S \times L \tag{2.4}$$

Therefore,

$$DE = 0.0287 \times 40 \text{ ft} = 1.15 \text{ ft}$$

The difference in elevation is then added to 44.37 to determine the desired spot elevation (Fig. 2.12).

$$44.37 + 1.15 = 45.52 \text{ ft}$$

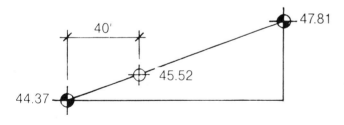

FIGURE 2.12. Section for Example 2.5.

SLOPES EXPRESSED AS RATIOS AND DEGREES

Often slopes are expressed as ratios such as 4:1. This means that for every 4 ft of horizontal distance there is a 1-ft vertical change either up or down. On construction drawings, particularly sections, ratios may be shown graphically using a triangle as illustrated in Fig. 2.13a. In expressing ratios, the horizontal number should always be placed first. Conversely, ratios may be expressed by their percentage equivalents. A 4:1 ratio is equivalent to a 25-ft vertical change in 100 ft, or a 25% slope.

Civil engineers and surveyors may express slope in degrees rather than percentages, although this terminology is rarely used by landscape architects. However, the conversion from one system to the other is quite simple, since the slope formula is a basic trigonometric function. The percentage of slope is actually the tangent of the angle of inclination, as illustrated in Fig. 2.13b. It should be noted that the rate of change between percentages and degrees is not arithmetically constant. For example, a 100% slope equals 45°, whereas a 50% slope equals 26°34′.

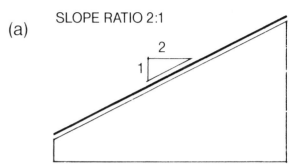

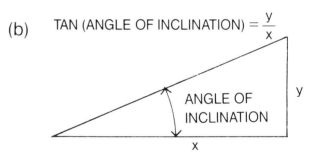

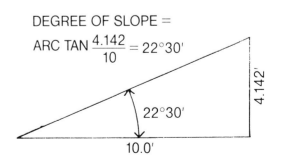

FIGURE 2.13. Alternative Methods of Expressing Slope.
 a. As a ratio.
 b. In degrees.

EXERCISES

2.1 At a scale of 1 in. = 100 ft plot the following spot elevations on a rectangular grid. The grid cells are 100 ft on each side. The coordinate letter indicates the row, while the coordinate number indicates the column.

A1 70.2, A2 69.5, A3 69.2, A4 68.0, A5 67.5, A6 65.1
B1 75.8, B2 75.8, B3 74.2, B4 73.8, B5 72.6, B6 73.1
C1 74.9, C2 74.5, C3 73.3, C4 72.1, C5 71.4, C6 70.5
D1 73.8, D2 73.6, D3 72.5, D4 71.9, D5 73.2, D6 72.4
E1 75.0, E2 74.1, E3 73.9, E4 73.8, E5 73.2, E6 72.9

2.2 Draw contour lines at a 1-ft vertical interval for the grid constructed in Exercise 2.1. Analyze the landform represented by the contours and list the landform features which are present.

2.3 From the scale and contour intervals shown in Fig. 2.14, interpolate the elevations for the points indicated on the contour plans.

2.4 At a 3.6% slope, how far downhill is the 47-ft contour line located from a 47.72 spot elevation?

2.5 Two trees are located 87 ft apart (measured horizontally) on a slope with a 0.04 ft/ft gradient. If the elevation of the higher tree is 956.58 ft, what is the elevation of the lower tree?

2.6 The drop-off point at a hospital entrance is 3.2 ft lower than the door elevation. How long must a ramp be to maintain an 8% slope? What is the length of the ramp measured *along the slope?*

2.7 Two drain inlets are located 140 ft apart, measured horizontally, on a road with a 0.053 gradient. If the elevation of the lower drain is 72.83, what is the elevation of the upper drain?

2.8 Convert the following slopes to ratios:
a. 5% b. 33% c. 50% d. 20%

2.9 Convert the slopes a through d in Exercise 2.8 to degrees and minutes.

(a)
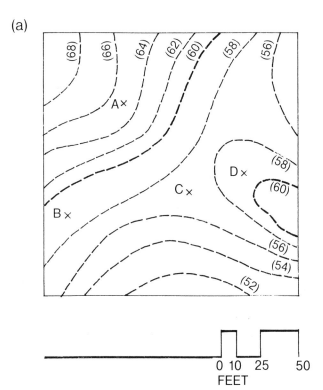

(b)
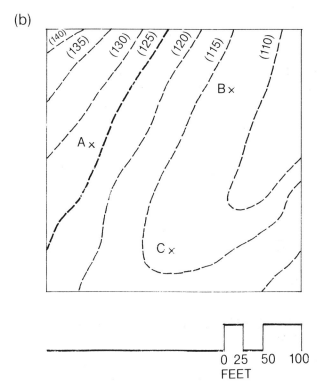

FIGURE 2.14. Contour Plans for Exercise 2.3.

3
Slope Formula Applications

The purpose of this chapter is to demonstrate, by example, the variety of uses of the slope formula. Several problems are presented; in addition to illustrating the mechanics of manipulating the formula, they also introduce concepts that are basic to the grading process.

SLOPE ANALYSIS

To determine the best areas for placing buildings, roads, parking lots, and other features on a particular site, landscape architects often conduct an analysis of the steepness of the terrain. This process, commonly referred to as *slope analysis,* provides information that can be used in conjunction with other considerations such as economics, vegetation, drainage, and soils, in making site planning decisions.

In order to conduct a slope analysis on a map, the following information is required: horizontal scale, contour interval, and percentage of slope categories. The scale and contour interval are established by the contour map used for the analysis. The slope categories are selected by the evaluator on the basis of the amount of change in elevation, the complexity of the landforms, and the types of activities to be accommodated. An example of slope categories can be found in the soil classification system of the United States Department of Agriculture (USDA) Soil Conservation Service for land use.

Example 3.1

The task for this problem is to analyze the slopes illustrated by the contour plan in Fig. 3.1. The scale is noted on the figure, and the contour interval is 2 ft. The slope categories are divided into three groups: less than or equal to 5%, greater than 5% but less than 15%, and equal to or greater than 15%. To execute the analysis, the allowable horizontal distance between contours for each slope category must be determined.

Solution. For the first category the range is from 0% to 5%. The unknown variable in the slope formula is the distance L, whereas the known variables are the difference in elevation (the contour interval, or 2 ft) and the slope, in this case the range from 0% to 5%. Substituting into the slope formula $L = DE/S$, at 0% the contours must be infinitely far apart; at 5% the distance between contours equals 40 ft.

$$L = \frac{2.0}{0.05} = 40 \text{ ft}$$

Therefore, any adjacent contours closer than 40 ft represent a slope greater than 5%.

For the next category, greater than 5% and less than 15%, the maximum distance between adjacent contours is the 40 ft already determined by the first category. The minimum distance, or the steepest slope, is determined by substituting the upper limits of the category (15%) into the formula.

$$L = \frac{2.0}{0.15} = 13.3 \text{ ft (or 13 ft)}$$

Thus, slopes ranging from greater than 5% to less than 15% are represented by contours which range from 13 to 40 ft apart. (Because of the small map scale the 13.3 ft may be rounded to 13 ft.)

The third category, slopes equal to or greater than 15%, is represented by adjacent contours that are not more than 13 ft apart.

A simple technique to determine the percentage of slope between contours is to construct a wedge-shaped piece of

18   Slope Formula Applications

FIGURE 3.1. Contour Plan.

paper indicating the critical dimensions between slope categories drawn to the proper map scale.

The wedge in Fig. 3.2 is a graphic interpretation of the information illustrated by the scales to the right. By moving the wedge over the contour map, it is easy to determine the distance between adjacent contours and the corresponding slope category. For ease of visualization and evaluation, the slope analysis is usually presented graphically on the topographic map as in Fig. 3.3. It should be noted that computers can be used very effectively to produce slope analysis maps.

SLOPES FOR SURFACE DRAINAGE

A primary objective in grading, in most instances, is to slope the ground surface away from buildings to ensure proper drainage of storm water. Storm water is directed away from buildings to prevent potential leakage into interior spaces; saturation of soils, which reduces bearing capacity; and potential adverse effects of moisture on building materials.

Example 3.2

To illustrate the concept of drainage away from a structure, often referred to as *positive drainage,* the rectangle in Fig. 3.4 represents a building measuring 30 ft by 50 ft. The objective of the problem is to locate and draw the 25- and 26-ft contour lines so that a 3% slope away from the building is achieved. The spot elevations at the exterior corners of the building are as indicated.

Solution.   The first step in the solution is to determine the distance from the spot elevations to the whole number contour lines. Along the south face of the building there is a change in elevation from 25.4 to 26.2. Thus a 26.0-ft spot elevation exists between these points and is easily located using the following proportion:

Slopes for Surface Drainage   19

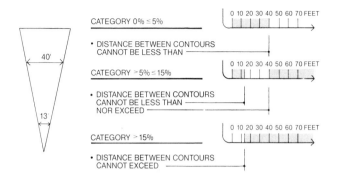

FIGURE 3.2. Wedge and Graphic Scales for Conducting Slope Analysis.

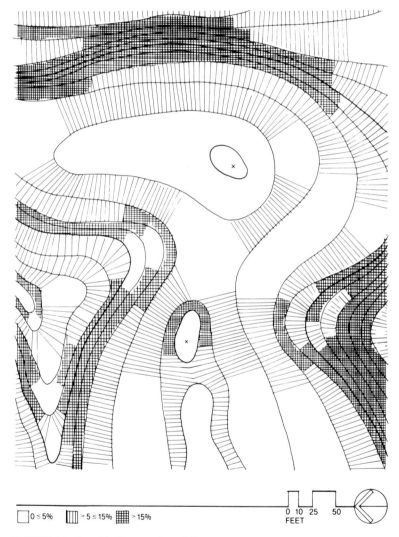

FIGURE 3.3. Graphic Presentation of Slope Analysis.

$$\frac{x \text{ (horizontal distance from 26.2 to 26.0)}}{50 \text{ (horizontal distance from 26.2 to 25.4)}} = \frac{0.2 \text{ (elevation change from 26.2 to 26.0)}}{0.8 \text{ (elevation change from 26.2 to 25.4)}}$$

$$\frac{x}{50} = \frac{0.2}{0.8}$$

$$x = \frac{0.2 \times 50}{0.8} = 12.5 \text{ ft}$$

Therefore, the 26.0-ft spot elevation is located 12.5 ft (or one-fourth the length of the south face) from the southeast corner.

The same process is applied to the east face, again to locate the 26.0-ft spot elevation.

$$\frac{x \text{ (horizontal distance from 25.9 to 26.0)}}{30 \text{ (horizontal distance from 25.9 to 26.2)}} = \frac{0.1 \text{ (elevation change from 25.9 to 26.0)}}{0.3 \text{ (elevation change from 25.9 to 26.2)}}$$

$$\frac{x}{30} = \frac{0.1}{0.3}$$

$$x = \frac{0.1 \times 30}{0.3} = 10 \text{ ft}$$

As a result, the 26.0-ft spot is located 10 ft, or one-third the distance, from the northeast corner.

The next step is to determine the distance at which the 26-ft contour line must be located in any direction from the southeast corner in order to fulfill the requirement of a 3% slope away from the building. The known values are $S = 0.03$ and $DE = 0.2$ ft, which may be substituted into the equation $L = DE/S$ or $L = 0.2/0.03$. The computed distance is 6.67 ft, or approximately 6.7 ft. An arc with a radius of 6.7 ft is constructed to the proper scale around the southeast corner and lines are drawn tangent to this arc from the previously determined 26.0-ft spot elevations. The 26-ft contour line is now complete.

The same method is used to compute the distance from each of the four corners to the 25-ft contour line. These computations are as follows:

Northeast corner: $\dfrac{0.9}{0.03} = 30$ ft

Southeast corner: $\dfrac{1.2}{0.03} = 40$ ft

Southwest corner: $\dfrac{0.4}{0.03} = 13.3$ ft

Northwest corner: $\dfrac{0.1}{0.03} = 3.3$ ft

Using these distances as radii, arcs are constructed around the appropriate corners and the 25-ft contour line is completed by connecting the arcs with tangent lines as illustrated in Fig. 3.4.

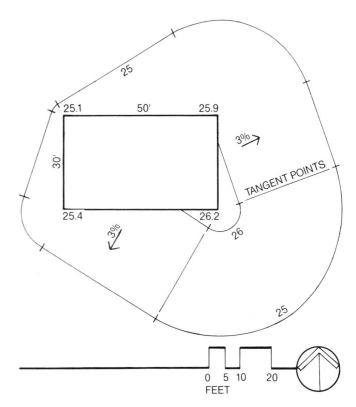

FIGURE 3.4. Plan for Example 3.2.

TERRACE GRADING

Another common grading problem is the construction of relatively level areas, or terraces, on sloping terrain. Most hillside development, whether it be for outdoor living areas, recreation facilities, or circulation systems, requires some form of terracing. In section, terraces may be graded in one of three ways: completely on fill, completely in cut, or partially on fill and partially in cut. *Fill* is soil that has been added to raise the elevation of the ground; *cut* is a surface from which soil has been removed to lower the ground elevation. For further discussion see Chapter 11, particularly Fig. 11.2. The three terrace conditions are illustrated in Fig. 3.5. Grading techniques for the first two conditions are explained in the next two examples.

Example 3.3 Terrace on Fill

A terrace measuring 25 ft by 40 ft is to be constructed with its south edge at elevation 220.0 as shown in Fig. 3.6. The terrace will slope downward at 3% toward the north for drainage and the side slopes will be graded at a ratio of 3 to 1 (3:1, or 1-ft drop in 3-ft horizontal distance). The purpose for making the side slopes quite steep is to return to the existing grade in the shortest distance possible. This reduces the amount of disturbance caused by grading and reduces cost. From the information provided, draw the proposed contour lines.

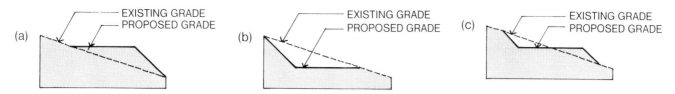

FIGURE 3.5. Terrace Sections. a. Terrace constructed on fill. b. Terrace constructed in cut. c. Terrace constructed partially on fill and partially in cut.

Solution. The first step in solving this problem is to determine the elevation along the north, or lower, edge of the terrace. From the equation $DE = S \times L$, the north edge is 0.03×25 ft $= 0.75$ ft lower than the south edge or 219.25 ft (220.0 − 0.75).

The next step is to determine the distance from the north edge of the terrace to the 219-ft contour line. Since the side slopes are to be graded at 3:1, the horizontal distance from the terrace edge to the 219-ft contour line is 0.75 ft as calculated by the following proportion:

$$\frac{x \text{ (horizontal distance from 219.25 to 219.0)}}{0.25 \text{ (elevation difference from 219.25 to 219.0)}} = \frac{3}{1} \text{ (proposed ratio)}$$

$$x = \frac{0.25 \times 3}{1} = 0.75 \text{ ft}$$

This distance is marked off along lines drawn perpendicular to the edges of the terrace at the northeast and northwest corners as shown in Fig. 3.6. (It is advisable to study Figs. 3.6 and 3.7 carefully while following the procedure outlined in the text.) From the point of the 219-ft spot elevation, the remaining whole number spot elevations (i.e., 218, 217, 216) can be located by progressing along the line in 3-ft increments, since for every 3 ft of horizontal distance there is a 1-ft vertical drop. These points are used for the construction of the proposed contour lines.

The same procedure is followed at the south edge of the terrace, where again lines are drawn perpendicular to the terrace at the southeast and southwest corners. Since the elevation of the south edge is already at a whole number (220), the remaining whole number spot elevations can be located by progressing out from the edge in 3-ft increments.

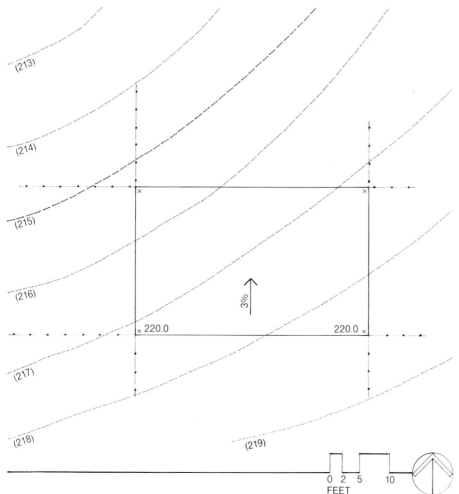

FIGURE 3.6. Plan for Example 3.3.

Beginning with the 219-ft contour line, draw straight lines through the 219-ft spot elevations until the lines of adjacent sides intersect. The proposed 219-ft contour line is a closed contour, since it never intersects the 219-ft contour line already existing on the site. Proceeding with the proposed 218-ft contour line and following the same technique, the new 218-ft contour line intersects the existing 218-ft contour line at two points. To continue the proposed contour line beyond the intersection point would result in cut that is unnecessary, since the terrace is entirely on fill. Therefore, the new contour lines are drawn only to the point where they intersect the corresponding existing contour lines. This is the point at which the grade returns to the original or existing ground surface. This is referred to as the *point of no fill* (or *no cut*). The procedure is continued for successively lower contour lines (217, 216, 215, etc.) until the point where existing contours are no longer disturbed is reached, as shown in Fig. 3.7.

Frequently the points of no fill (or cut) are delineated by a line which also serves as the limit line for the grading contractor's work. Where there is no intersection of existing and proposed contour lines of equal elevation to delineate the no fill (or cut) line easily, a section showing the proposed and existing grade lines may be constructed. The point of no fill (or cut) occurs where the two grade lines intersect.

With regard to the shape of the side slopes constructed in this example, two points must be made. First, the side slopes in plan view form planes with distinct intersections. This is difficult to construct and maintain and, within a natural context, usually does not blend well with the surrounding landscape. For these reasons, the contours are given a smoother and more rounded appearance as shown in Fig. 3.8.

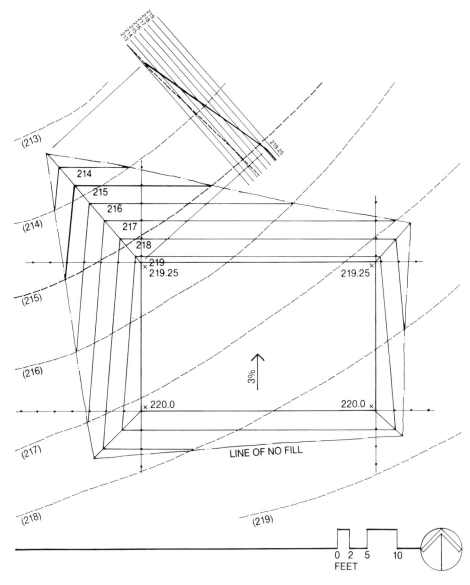

FIGURE 3.7. Completed Contour Plan. The section is used to determine the location of the no cut–no fill line between contour lines 214 and 213.

Terrace Grading 23

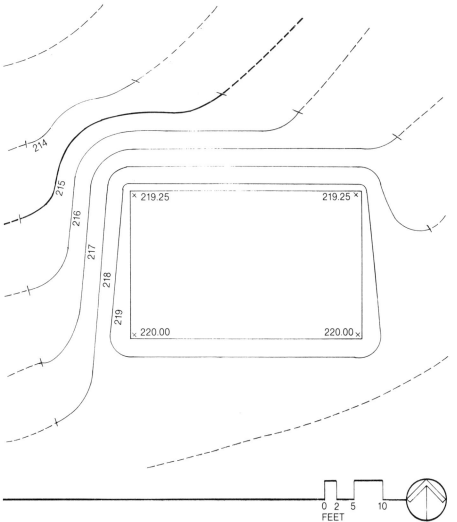

FIGURE 3.8. Contour Lines Adjusted to Provide Smoother, More Rounded Appearance.

The second point pertains to the relationship of the rather steep side slopes to the edge of the terrace and where the side slopes meet existing grade. As constructed, there is an abrupt change in grade at the top and bottom of slope as shown in section in Fig. 3.9a. Again, not only is this condition difficult to maintain, it is also subject to erosion, particularly at the top of the slope. An alternative is to provide additional space at the top and bottom of the slope to allow for a smoother transition as illustrated in Fig. 3.9b.

Example 3.4 Terrace in Cut

A terrace measuring 25 ft by 30 ft is to be built with the southern, or higher, edge at elevation 108.0 ft as shown in Fig. 3.10. The surface of the terrace will slope toward the north at 2%. The proposed side slopes, which are in cut, will have a 3 to 1 slope. From the information given, locate the proposed contour lines.

Solution. The procedure for solving this problem is similar to that in the previous example. First, the elevation of the northern edge of the terrace is determined to be 107.5 ft (0.02 × 25 = 0.5 ft and 108.0 − 0.5 = 107.5 ft). From analyzing the relationship between the existing and proposed elevations, grading essentially is not required

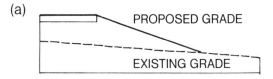

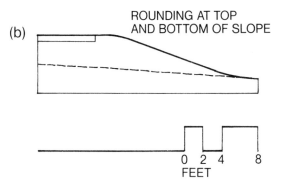

FIGURE 3.9. Slope Sections. a. Abrupt transition at top and bottom of slope. b. Rounded transition at top and bottom of slope. Note that providing a transition area requires additional horizontal distance.

along the northern edge, since the proposed and existing grades correspond. Along the southern edge there is 1.5 to 2.0 ft of cut. This means that the ground must slope uphill from this edge at the proposed 3:1 ratio in order to return to the existing or original grade. The solution is shown in Fig. 3.11. The completed contour plan in Fig. 3.11 contains the same problems created by intersecting planes which were discussed in relationship to Figs. 3.8 and 3.9, and the same principles should be applied in refining this proposed grading plan.

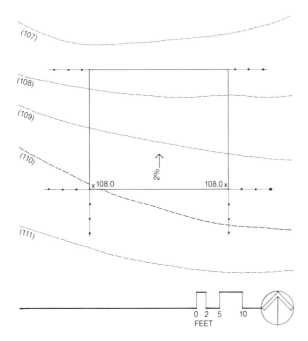

FIGURE 3.10. Plan for Example 3.4.

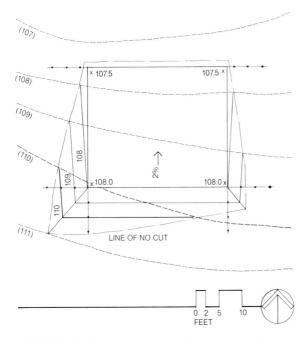

FIGURE 3.11. Completed Contour Plan Based on Criteria.

PATH LAYOUT WITH A MAXIMUM GRADIENT

Often the design of roads, walks, and paths may not exceed a specified gradient or slope. Slope criteria may be established by building and zoning codes, highway departments, accessibility for the handicapped, or common-sense constraints such as climatic conditions and age of anticipated users. The following example illustrates, within a natural landscape, how a path may be designed not to exceed a specified gradient, while minimizing the amount of grading required.

Example 3.5

In Fig. 3.12 the objective is to construct a path from the road pull-off to the lake dock at a desired maximum slope of 4%. Both the horizontal scale of the plan and the contour interval must be known in order to proceed with the problem. In this example the scale is shown on the figure, and the contour interval is 1 ft. The path begins at point A along the 238-ft contour line.

Solution. First, the known values must be substituted into the equation $L = DE/S$ to determine the length of the path between adjacent contours in order to satisfy the 4% criterion.

$$L = \frac{1.0}{0.04} = 25 \text{ ft}$$

By drawing an arc with a radius of 25 ft at the proper scale with point A as its center, two intersection points are obtained along the 237-ft contour line. A line drawn from point A to either of these two points will have a gradient of 4%. Any line drawn within the shaded triangle formed by the two path alternatives will be shorter than 25 ft and therefore steeper than 4%. Conversely, lines drawn beyond the boundaries of the triangle will be less than 4%. Selecting one of the previously established points on the 237-ft contour line as the center point, another arc is drawn with a radius of 25 ft. This arc intersects the 236-ft contour line at two points. Selection of path direction is based on a number of considerations: overall design intent, views, location of trees, soil stability, and so forth. Progressing downhill on a contour-by-contour basis, the entire alignment for the path sloping at 4% may be established.

An alternative to the process described is to take the difference in elevation between point A and the lake dock. By dividing by the desired percentage of slope, the total length of the path is determined. Assuming constant gradients, a path of shorter length will exceed 4%, and a longer path will be less than 4%.

$$238 - 229 = 9 \text{ ft}$$

$$\frac{9}{0.04} = 225 \text{ ft}$$

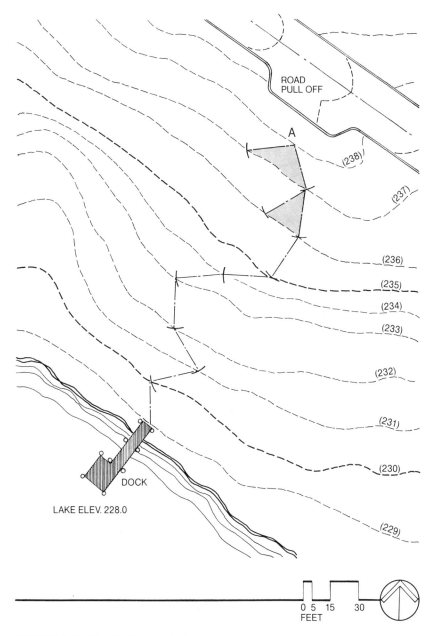

FIGURE 3.12. Plan for Example 3.5.

Thus, any combination of curves and straight lines with a total length of 225 ft and a constant gradient will meet the desired 4% criterion. However, this method does not necessarily relate as well to the natural topography of the land as the first method does.

GRADING OF ROADS

A road represents a microcosm of most grading problems found in landscape architecture. These include ridges (crowns), valleys (swales), slopes in two directions (crosspitch), and vertical planes (curbs). Developing the ability to grade a road and to visualize this process in three dimensions will measurably improve one's understanding of manipulating contours. However, before discussing how to grade a road, it is necessary to define the terminology associated with road construction.

Definitions

Crown
Crown is the difference in elevation between the edge and the center line of a roadway. The primary purpose of the crown is to increase the speed of storm runoff from the road surface. A secondary purpose is visually to separate opposing lanes of traffic. Crown height may be expressed in inches or inches per foot. In the latter case, the total crown height is calculated by multiplying half the road width by the rate of change. For example, the crown height for a 24-ft-wide road with a crown of ¼ in./ft is 3 in. (12 ft × ¼ in./ft = 3 in.) There are three basic types of road crown section (Fig. 3.13).

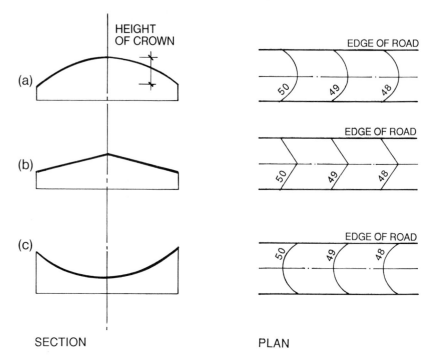

FIGURE 3.13. Road Crowns. a. Parabolic section. b. Tangential section. c. Reverse crown.

Parabolic Section

A parabolic crown is commonly used in asphalt construction. The change in slope direction at the roadway center line is achieved by a rounded transition. Contour lines point in the downhill direction (similar to a ridge contour signature).

Tangential Section

A tangential crown is most often found with concrete surfaces, since it is easier to form. The center line of the roadway is visually emphasized because of the intersection of the sloping planes along this line. Again the contour lines point in the downhill direction, but they are V-shaped rather than rounded in appearance.

Reverse Crown

A reverse crown may be either parabolic or tangential in section. It is typically used where it is not desirable to direct storm runoff to the edge of the road or in restricted conditions such as urban alleys. Its contour signature is similar to that of a valley.

It should be noted that not all roads have crowns. Some roads are cross-sloped: in other words, storm runoff is directed from one side of the road to the other. This type of road section is also used to bank road curves to counteract overturning forces (see Chapter 12).

Curb

A curb is a vertical separation at the edge of the roadway. It is usually 6 in. high but may range from as low as 4 in. to as high as 8 in. Curbs are used to direct and restrict storm runoff and to provide safety for pedestrians along the road edge. In addition to vertically faced curbs, beveled or rounded cross sections may be used (Fig. 3.14).

Swale

A swale is a constructed or natural drainage channel which has a vegetated surface (usually grass). A *gutter* is a paved swale. The depth of swales (but not necessarily gutters) is usually measured as the difference in elevation between

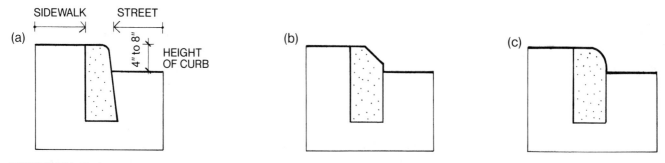

FIGURE 3.14. Curbs. a. Batter-faced section used for typical street curb. b. Beveled section. c. Rounded section. Both b and c are referred to as mountable curbs.

the center line and a point at the edge of the swale on a line taken perpendicular to the center line (Fig. 3.15). Since swales are depressions similar in form to valleys, the contour signatures are also similar. Swales are commonly used to intercept, direct, and control storm runoff, which is discussed in Chapter 10.

However, it is important to understand how swales are represented by contour lines. Figure 3.16 illustrates three swales of different depths but with the *same* width and longitudinal gradient. Often the high point of a swale is the dividing point for two swales sloping in opposite directions. This point is referred to as a *saddle*, since it is simultaneously a high point (parallel to the direction of the swales) and a low point (perpendicular to the direction of the swales). This phenomenon is illustrated in Fig. 3.17.

Example 3.6

Figure 3.18 illustrates, in plan and section, a variety of conditions found in conjunction with roads and streets, including crown, curb, sidewalk, and swale. From the established spot elevation of 25.42 ft at point *A* the location of the 25-ft contour line is to be determined on the basis of the following criteria: a 6-in. crown height, a 6-in. curb height, a 4-in. swale depth, a 3% slope parallel to the direction of the street, and a 2% slope across the sidewalk perpendicular to and downward toward the street. The process of locating the 25-ft contour line is explained in this example.

Solution. The first step is to locate the 25.0-ft spot elevation along the road center line. Since the difference in elevation is 0.42 ft and the slope is given at 3%, the distance between 25.42 and the 25.0 spot elevations equals 14.0 ft (0.42/0.03 = 14.0 ft). In terms of the stated criteria, the crown height is 6 in. Therefore, the spot elevation at the edge of the road opposite the 25.0-ft spot elevation is 6 in. (0.5 ft) lower than the center line, or 24.5 ft. The 25.0-ft spot elevation at the edge of the road is located uphill from this point. The difference in elevation is 0.5 ft; therefore, the distance from 24.5 to 25.0 is 16.7 (0.50/0.03 = 16.7 ft).

On the east side of the road, the edge is a 6-in.-high curb. This means that the elevation at the top of the curb is always 0.5 ft above the elevation of the edge of the road pavement. Where the edge of the road is elevation 25.0 ft, the top of the curb is 25.5, and where the road is 24.5 ft, the top of the curb is 25.0. Therefore, the 25.0 contour line follows along the face of the curb until it emerges at the 25.0-ft spot elevation as illustrated in plan (Fig. 3.18) and in isometric (Fig. 3.19).

Adjacent to the curb is a sidewalk that slopes toward the road at 2%, while sloping 3% parallel to the direction of the road. As a result, the far edge of the sidewalk is higher than the edge along the curb. The difference in elevation between the two edges of the walk is the width, 6 ft, multiplied by the slope, 2%, or 0.12 ft. Thus, the elevation at the far edge of the sidewalk directly opposite the 25.0-ft spot elevation at the top of the curb is 25.12 ft. The 25.0-ft spot elevation along this edge is located

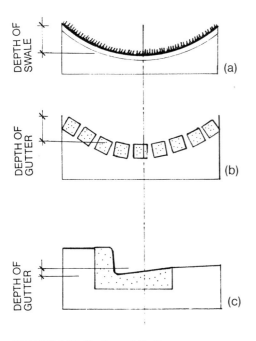

FIGURE 3.15. Swales and Gutters.
a. Vegetated parabolic swale.
b. Paved gutter.
c. Combination curb and gutter.

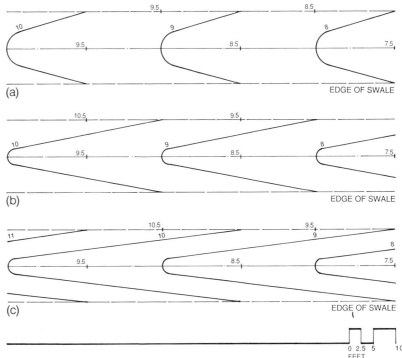

FIGURE 3.16. Plan of Three Swales with Different Depths but the Same Gradient (3%) and Width (15 ft). a. 6 in. deep. b. 12 in. deep. c. 18 in. deep.

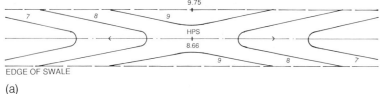

(a)

FIGURE 3.17. Saddle Created by the High Point of Two Swales Sloping in Opposite Directions. a. Plan. b. Axonometric.

downhill parallel to the direction of the road at a slope of 0.03. The distance is computed by dividing the difference in elevation, 0.12 ft, by the slope 3% (0.12/0.03 = 4.0 ft). Therefore, the 25.0 spot elevation is located 4.0 ft from the 25.12-ft spot elevation in the downhill direction. The 25-ft contour line is constructed across the sidewalk by connecting the 25.0-ft spot elevation at the top of the curb with the 25.0-ft spot elevation at the edge of the sidewalk.

On the left side of the road is a swale that is 6 ft wide and 4 in. deep. Therefore, the center line of the swale is 4 in., or 0.33 ft, lower than the two edges. As a result, where the edge of the road is at elevation 25.0 ft, the elevation at the center line of the swale is 24.67 ft. The 25.0 spot elevation along the swale center line is located by dividing the difference in elevation, 0.33 ft, by the slope of the swale center line, which is the same as that of the road, 3% (0.33/0.03 = 11.0 ft). Measuring 11 ft uphill along the center line from the 24.67-ft spot elevation locates the 25.0-ft spot elevation. The 25-ft contour line is constructed as shown in Fig. 3.18. Note that the crown of the road and the bottom of the swale are indicated by rounded contours which reflect the rounded shape of these elements in section, and both are shown symmetrical about their respective center lines.

Grading by Proportion

Grading a road provides an excellent opportunity for illustrating the use of visual and graphic techniques to locate proposed contour lines. The key to this process is to remember that there is a direct proportional relationship between change in elevation and horizontal distance between contour lines. This is demonstrated in Example 3.6, where a 1.0-ft change in elevation at a 3% slope requires a distance of 33.3 ft. For a 0.5-ft change in elevation only one-half that distance, or 16.7 ft, is required, whereas a 0.33-ft change requires one-third the distance, or approximately 11 ft.

Figure 3.20 outlines a graphic technique which may be used to lay out the contour lines for a symmetrical road crown. Once this technique is fully understood it can applied to *any* grading situation and can facilitate a quick visual approach to solving grading problems.

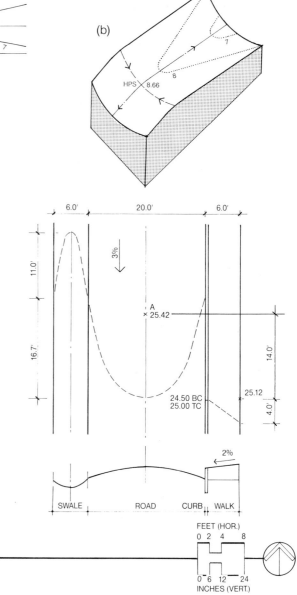

FIGURE 3.18. Plan and Section for Example 3.6.

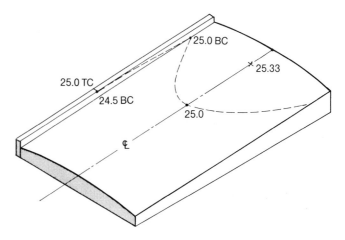

FIGURE 3.19. Path of Contour Line along Face of Curb.

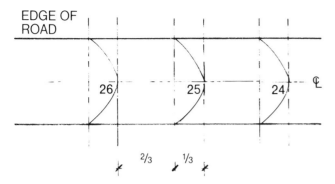

FIGURE 3.20. Graphic Technique for Establishing Contour Lines for Crowned Roads.
a. Locate whole number spot elevations along road center line.
b. Express crown height as fraction of a foot (e.g., 6 in. = 1/2 ft, 4 in. = 1/3 ft).
c. Divide space between whole spot elevations according to fraction.
d. Draw smooth curve from spot elevation to point where fraction lines cross the edge of the road.

Example 3.7

The final example of this chapter combines several of the points made in previous examples. For this problem only a portion of a proposed road is illustrated in Fig. 3.21. The objective of the problem is to locate and draw the proposed contour lines with a 1-ft interval according to the given criteria and to connect the corresponding proposed and existing contour lines where appropriate. The design criteria are as follows: a 4% slope toward the south parallel to the direction of the road; a 4-in. crown height; a 6-ft-wide swale along the west edge of the road, 6 in. deep and also sloping at 4%; a 6-ft-wide shoulder along the east edge that slopes at 2% perpendicular to and away from the road; and side slopes perpendicular to the center line of the road at a ratio of 3 to 1 in cut and fill. The spot elevation at point A has been established as 69.0 ft.

Solution. The first step is to locate the whole number spot elevations along the road center line. From the equation $L = DE/S$ the distance required for a 1.0-ft change in elevation at 4% is 25 ft. Spot elevations 68.0, 67.0, and 66.0 are marked off along the center line at 25-ft intervals. Because of the crown, points on both edges of the road on a line perpendicular to the center line will be 4 in., or 0.33 ft, lower than a point at the center. Thus, each whole number spot elevation at the edge will be located 8.25 ft in the uphill direction from the same elevation at the center of the road. This distance was calculated by dividing the difference in elevation, 0.33 ft, by the slope, 4% (0.33/0.04 = 8.25 ft). From this information spot elevations 69.0, 68.0, 67.0, and 66.0 can be located on both edges of the road.

By examining the existing grades and the proposed spot elevations, it is determined that the west edge of the road is in cut and the east edge is in fill. The swale that has been proposed along the west edge is necessary in order to intercept the storm runoff from the cut slope. The swale is 6 in. (0.50 ft) deep, which means that a point on the center line of the swale is 0.50 ft lower than the edge of the road. Each whole foot spot elevation along the swale center line is 12.5 ft uphill from the corresponding elevation at the road edge (0.50/0.04 = 12.5 ft). The whole foot spot elevation along the far edge of the swale is opposite the whole foot spot elevation at the edge of the road.

A 6-ft-wide shoulder, which slopes away from the road, is proposed along the east edge. The far edge of the shoulder is 0.12 ft *lower* than the edge of the road. This is determined by multiplying the slope by the shoulder width (0.02 × 6 ft = 0.12 ft). Again, the whole foot spot elevation at the edge of the shoulder must be located in the uphill direction from the whole foot spot elevation at the edge of the road. The distance uphill is calculated by dividing the difference in elevation, 0.12 ft, by the slope, 4%, which equals 3.0 ft.

The 3 to 1 side slopes begin at the outside edges of the swale and the shoulder. Using a method similar to that in Example 3.3, lines are drawn perpendicular to the road center line through the full foot spot elevations at the outside edges of the swale and the shoulder. Beginning at the respective edges, points are located at 3-ft intervals along the lines, since each 3-ft horizontal distance results in a 1-ft change in elevation. The proposed contour lines are located by connecting all the points of equal elevation determined in each of the previous steps. The proposed contour lines are drawn to the point where they intersect the corresponding existing contour lines. This is the point of no cut or fill, since beyond this point the existing contour line does not change.

VISUALIZATION OF TOPOGRAPHY FROM CONTOUR LINES

Occasionally, it may be somewhat confusing to interpret contour maps or to construct new contour lines for proposed grading projects. One of the most common problems among students is drawing contour lines in the proper direction for proposed crowns, curbs, and swales. The following simple technique provides a visual aid in these situations.

By turning the plan so that the viewer is looking in the downhill direction (i.e., the higher elevations are closer to the eye than the lower elevations), the proposed contour lines if drawn correctly resemble the proposed profile: that is, a cross section with an exaggerated vertical scale. Using Fig. 3.21 as an example, turn the page so that the plan is viewed from the north. Following the proposed 67-ft contour line, it can be seen that the contour plan of each of the proposed components reflects its respective cross section: the swale looks like a valley, the crown looks like a ridge, and the shoulder slopes away from the road. The existing contour slopes down toward the swale on the right (west), indicating cut, and slopes away from the shoulder on the left (east), indicating fill.

30 Slope Formula Applications

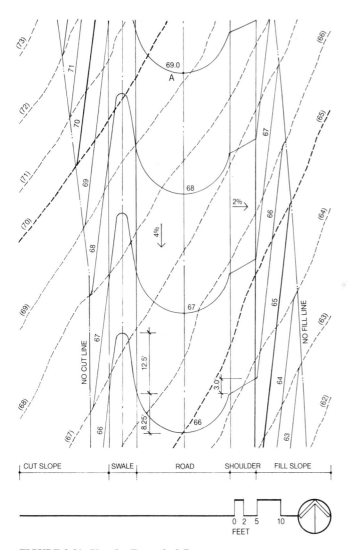

FIGURE 3.21. Plan for Example 3.7.

EXERCISES

3.1 In each of the two drawings (Fig 3.22), a plane sloping at 2% has been placed on existing topography. Provide the missing spot elevations, plot all proposed contour lines, and connect them to the existing contour lines. All cut and fill side slopes are to be graded at 4 to 1.

3.2 Draw the proposed contour lines at 1-ft intervals for the four road conditions illustrated in Fig. 3.23.

3.3 For the two road plans provided, draw the proposed contour lines at 1-ft intervals according to the criteria given and connect the proposed and existing contour lines where appropriate (Fig. 3.24a. & b.).

Visualization of Topography from Contour Lines 31

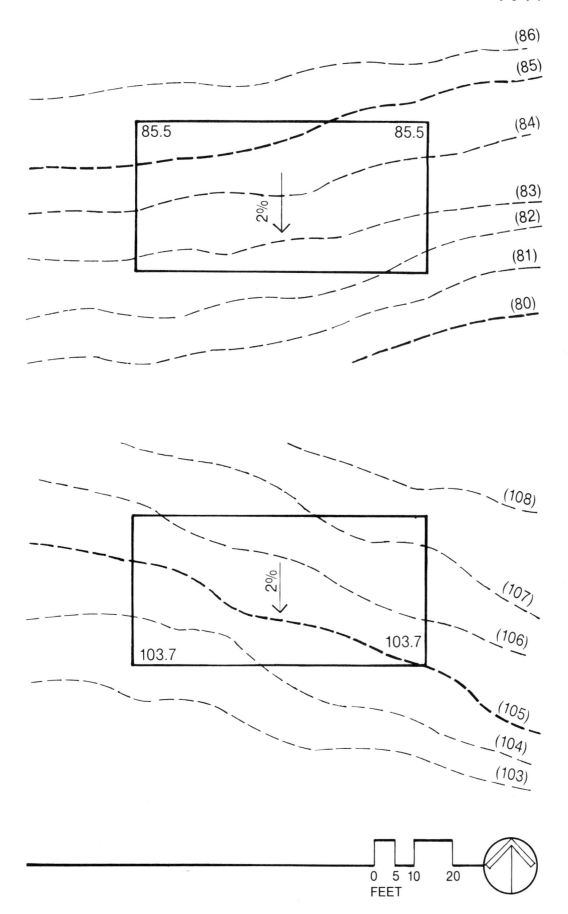

FIGURE 3.22. Plans for Exercise 3.1.

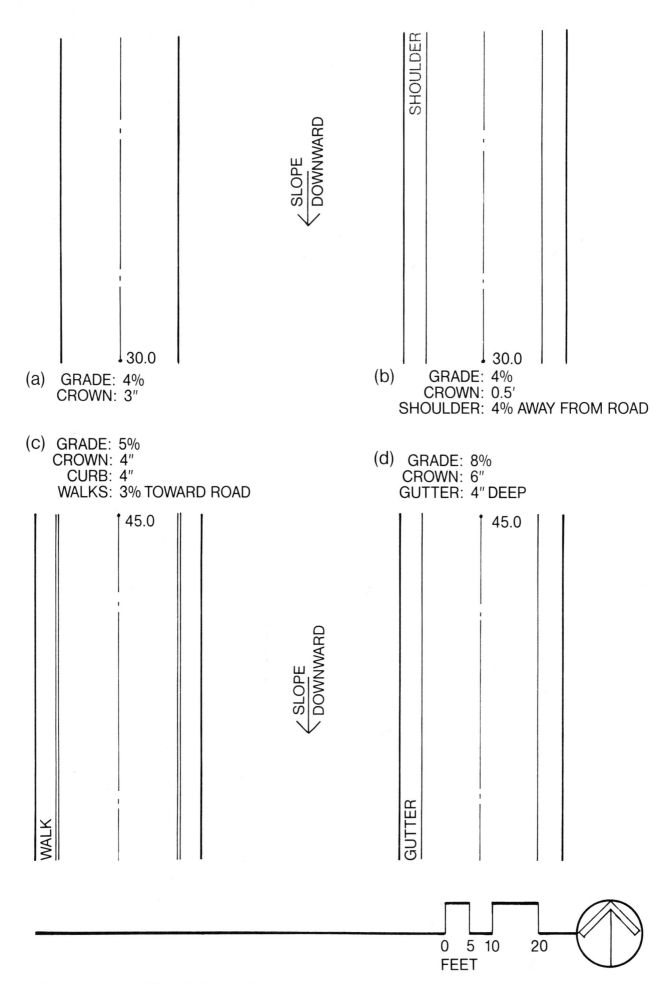

FIGURE 3.23. Plans and Criteria for Exercise 3.2.

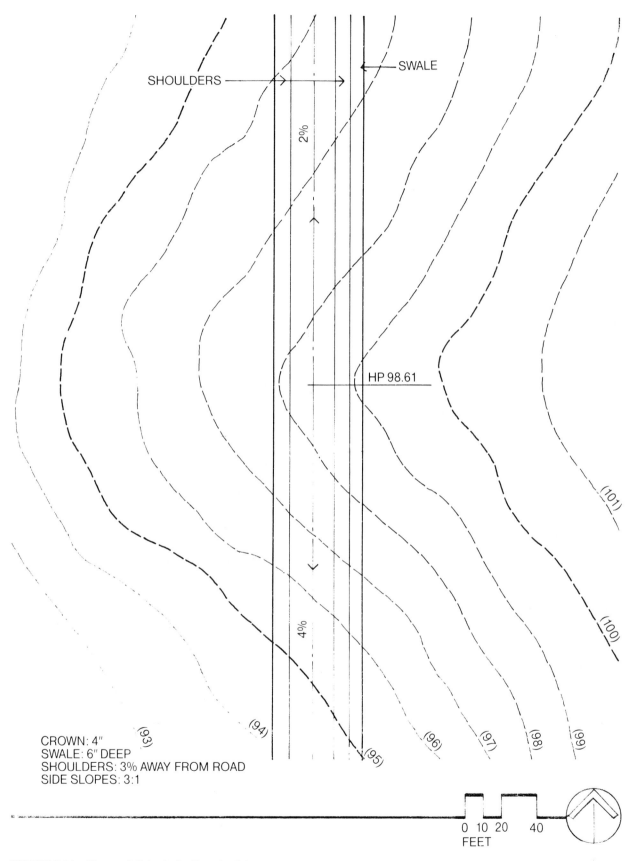

FIGURE 3.24a. Plans and Criteria for Exercise 3.3.

34 Slope Formula Applications

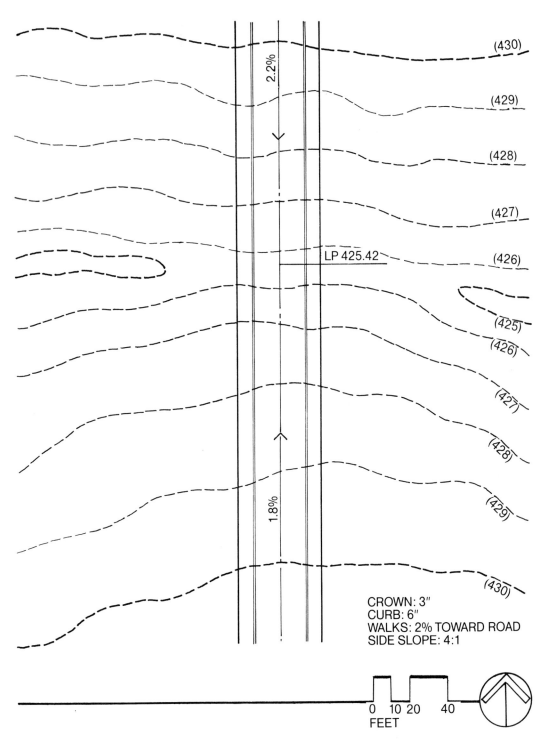

FIGURE 3.24b.

4
Grading Constraints

The problems presented in the last chapter were highly structured, since all grading criteria, such as proposed spot elevations, slopes, and crown and curb heights, were provided. However, grading is more than a mechanical process dictated by a predetermined set of criteria. Designing an appropriate grading scheme is a decision-making process based on a variety of constraints and opportunities.

The grading process must be viewed as a response to program concerns, design intent, and contextual conditions. The latter consist primarily of the existing physical conditions and the regulations that govern the development of a site. These establish the framework within which the program and design concept may be formulated. In turn, all three elements (program, site, and intent) create constraints that guide the decision-making process.

Although difficult to classify neatly, constraints constitute two broad categories: environmental and functional. Environmental constraints are those that deal with the natural conditions of the site. Functional constraints relate to the requirements of the activities that must be accommodated and other restrictions, including laws and regulations such as building and zoning codes. The following discussion examines each of the categories in more detail.

ENVIRONMENTAL CONSTRAINTS

Topography

The most obvious constraint is the existing topography. The existing landform should be analyzed carefully in order to guide the design of the proposed development. High and low points should be located, percentages of slope inventoried, and extent of relatively level and steep areas determined. Coordinating the proposed grading with the existing conditions, in most cases, reduces development costs and results in a more desirable final product.

Drainage

It is difficult to separate the act of grading from the act of accommodating and controlling storm water runoff, since one directly affects the other. This section briefly discusses storm drainage as a grading constraint. A more complete discussion of storm runoff and drainage systems is presented in Chapters 7 through 10.

A good rule of thumb to follow in order to reduce impacts and potential problems is to conform as closely as possible to the established natural, as well as constructed, drainage pattern within the proposed development. Before conforming, however, it is important to evaluate the existing system to determine whether it is functioning correctly both ecologically and hydraulically and whether it has the capacity to absorb any anticipated changes. Hydrologic problems that result from changes in the character of the environment have been well documented. Increased use of impervious surfaces such as pavements and roofs, channelization of streams, and floodplain encroachment, particularly in urban and suburban areas, have lowered water tables; increased the fluctuation of water levels in streams, ponds, and wetlands; and increased potential flood hazards. The impacts of these problems are also obvious. Flooding is a threat to safety, health, and personal well-being; lowering the water table directly affects water supply. Fluctuation of water levels will have an impact on plant and animal habitats, since such conditions require a high degree of tolerance to ensure survival. Acknowledging and understanding these problems will lead to a more sensitive handling of storm water runoff.

Vegetation

Disturbing the earth around existing vegetation generally has a detrimental effect on the health of plants. The impact is often not readily apparent. There have been instances in which architects and landscape architects have developed areas quite close to specimen trees. The photographs taken immediately after completion of the project provided an excellent and sensitive impression, but the reality of the design became apparent 2 or 3 years later when the dead trees were removed. Although not necessarily assuring survival, protection of existing plant material is best accomplished by avoiding grade changes within the drip line of the plants. Cutting soil within the drip line removes surface roots, and filling within the same area reduces the amount of air available to the root system. The consequence of any changes will depend on the plant species and soil conditions.

Protective systems, as illustrated in Fig. 4.1, may be used where it is necessary to change the elevation around existing vegetation, particularly trees. However, these systems increase the cost of a project and still do not guarantee plant survival. A better approach is to develop a grading system that minimizes these types of disturbances.

Existing vegetation may also be affected by changes in drainage patterns. Directing storm runoff to areas that are normally dry, draining areas that are usually wet, or lowering the water table alter the environmental conditions to which plants have adapted. These alterations may cause severe physiological stress and result in a change in the plant community that inhabits the area.

Soils

In any building process, it is important to understand the nature of the construction material used. In the case of grading, the construction material is soil. For landscape architects some knowledge of its engineering, as well as its edaphological and pedological properties, is necessary, since soil is used as both building material and growing medium. Properties such as bearing capacity, angle of repose, shear strength, permeability, erodibility, frost action potential, pH level, and organic content establish the capabilities and limitations of a soil. As part of the preliminary planning process for any major project, an investigation of the soil and geologic conditions must be conducted and an engineering report prepared by qualified professionals. Such a report will help determine the fea-

FIGURE 4.1. Grade Changes at Existing Trees. a. For cut condition either a slope or retaining wall can be used beyond the drip line to attain the desired grades. b. In fill situations a retaining wall can be placed beyond the drip line or within the drip line if proper aeration measures, such as gravel vents, are provided. c. A wood retaining wall used to protect a tree in a fill condition. d. A concrete retaining wall used to protect a tree in a cut condition.

sibility of a project and will form the basis for structural design decisions. Even for small-scale projects, it is advisable to consult a soils or foundation engineer, particularly if poor or unfamiliar soil conditions are anticipated.

Bearing Capacity

Bearing capacity is the size of the load, usually expressed as pounds per square foot, which a soil is able to support. Hard, sound rock has the highest bearing capacity; saturated and organic soils have the lowest. If the bearing capacity of an existing soil cannot support the proposed load or structure, the soil must be removed and replaced with suitable material, or other engineering measures must be taken. Such measures employ piles, spread footings, floating slabs, and other structural techniques.

Angle of Repose

Angle of repose is the angle at which an unconsolidated soil naturally slopes. The angle varies with the soil grain size, grain shape, and moisture content. To maintain stability, cut or fill slopes should not exceed this angle. If they do, slippage of the soil may occur.

Permeability

Permeability pertains to the ease with which water can flow through soil or any other material. The permeability of a soil is one of the factors that directly affects the design of a storm drainage system. Where permeability is high, and the infiltration capacity of the soil is sufficient, water easily penetrates the soil and the amount of storm runoff is relatively low unless the soil is saturated. Where permeability is low, the volume of runoff is higher and the design of the drainage system becomes more critical. Ecologically, it is good practice to maintain as much permeable surface on a site as possible. This not only maintains the level of the water table but also normally results in lower site development costs.

Soil permeability also affects the location and design of septic systems and the need for, and method of, dewatering excavations during construction.

Shear Strength

Shear strength determines the stability of a soil and its ability to resist failure under loading. Shear strength is a result of internal friction and cohesion. *Internal friction* is the resistance to sliding between soil particles; *cohesion* is the mutual attraction between particles, which is due to moisture content and molecular forces. Under typical conditions, sand and gravel are considered cohesionless. Clay soils, on the other hand, have high cohesion but little internal friction. As a general rule, the slopes for cohesive soils require flatter angles as the height of the slope is increased. Because of their internal friction, the shear strength of sand and gravel increases in relation to increased normal pressure; therefore, the angle of slope does not decrease with increased height.

Slope failure occurs when shear stress exceeds shear strength. The reason for failure is either increased stress or decreased strength brought about naturally or induced by human activity. Examples of increased stress are increasing the load at the top of a slope or removing the lateral support at the base of a slope through excavation or erosion. Decreased strength, as well as increased stress, occur when the moisture content of the soil is increased. In conclusion, care must be taken when developing at the top or bottom of relatively large slopes, and particular attention must be given to the handling of storm water runoff.

Erodibility

All soils are susceptible to erosion. This is especially true during the construction process, since stabilizing surface material such as vegetation is removed and soils become unconsolidated as a result of excavating and scraping. Erosion not only results in the loss of soil but also causes problems by depositing the soil in undesirable places such as lakes, ponds, and catch basins. Many communities now require the submission of a soil erosion and sedimentation control plan before the start of construction. The plan indicates the temporary control measures to be taken during construction and often includes the permanent measures that will remain once construction has been completed. An extremely important step is to strip and stockpile the existing topsoil layer properly in order to prevent unnecessary loss of this important resource. After regrading has been completed, the topsoil can be replaced on the site. Erosion and sediment control are discussed further in Chapter 7.

Topographic factors influencing erosion are the degree and length of slopes. The erosion potential increases as either or both of these factors increase. Knowing the erosion tendency of a soil, therefore, will influence the proposed grading design by limiting the length and degree of slope.

Frost Action

In northern climates silty soils and soils with a wide, fairly evenly distributed range of particle sizes, referred to as *well-graded soils,* are subject to frost action. Soils that exhibit these characteristics will influence footing and foundation design for structures and base course design for pavements. Two problems associated with frost action are soil expansion, which occurs during freezing, and soil saturation, which occurs during thawing.

Organic and Nutrient Content and pH Level

Tests to determine the nutrient and organic content and pH are performed to assess the ability of a soil to support plant growth, since slope stabilization is often achieved through vegetative means. If the chemical condition of the soil prevents vegetation from becoming established, erosion will occur. By knowing the pH, nutrient level, and amount of organic matter, corrective steps can easily be taken to improve the growing environment.

FUNCTIONAL CONSTRAINTS

Functional constraints have been broadly interpreted to include not only the limitations resulting from the uses that must be accommodated but factors such as maintenance, economics, and existing restrictive conditions.

Restrictive Conditions

The first set of functional constraints is determined by legal controls and physical limitations. Property lines establish legal boundaries beyond which a property owner does not have the right to modify or change existing conditions. Therefore, grades along property lines cannot be altered and thus become a controlling factor. In urban situations where land is expensive and sites are relatively small, costly grade change devices may be necessary along property lines to maximize the usability and development potential of a site.

In addition to prohibiting physical change, it is usually illegal to increase the rate of flow of storm water runoff from one property to adjacent properties. To be in compliance, the rate of runoff after completion of construction must not exceed the rate before construction.

Most design projects are under the jurisdiction of a political authority, whether it be at the municipal, state, or federal level, and usually must comply with a variety of building and zoning code regulations. In order to obtain permits, approvals, and certificates of occupancy, it is necessary to meet the criteria and standards set by these regulatory authorities. Items that typically may be regulated include maximum allowable cross-slope for public sidewalks, maximum height of street curbs, acceptable riser/tread ratios for stairs, maximum number of stair risers without an intermediate landing, maximum slope for handicapped ramps, slope protection, drainage channel stabilization, retention ponds, and vegetative cover. Before starting a project, the designer should become familiar with all applicable regulations.

Existing as well as proposed structures establish controls that affect grading design. Door entrances and finished floor elevations must be met, window openings may limit the height of grades, and structural conditions may limit the amount of soil that can be retained or backfilled against a building. The ability to waterproof or not waterproof a particular structure may determine whether soil can be placed against it. The response to these various conditions will influence the form of the grading design.

Utility systems, particularly those that flow by gravity or gravity-induced pressure, such as storm drainage, domestic water supply, and sanitary sewers, establish additional criteria that may control the grading process. If a proposed storm drainage system is connected to an existing system, then obviously the elevation of the existing system cannot be higher than that of the proposed system at the point of connection; if it is, pumping stations will be required. To prevent this from occurring, it is wise to work back from the known elevation at the outlet when establishing grades along the new drainage system. Also, the capacity of the existing system, including pipes and swales, must be evaluated to determine whether additional storm runoff can be accommodated. Domestic water and sanitary sewer connections also may require pumping if the required flow cannot be achieved naturally by gravitational pressure.

Activities and Uses

The two most prevalent uses that must be accommodated by grading are buildings and circulation systems. Typically, these create the most difficult problems and the most limiting constraints that must be dealt with in the grading process. In a holistic sense, the placement of buildings in the landscape and the resultant pedestrian and vehicular access patterns should work in concert with each other as well as with the natural and cultural context in which they are placed. This has not always been the case.

Quite often the decision to locate a building is based on a singular criterion such as zoning setbacks, initial cost, or view from the site, without a complete understanding of how this placement will affect the construction of roads, parking areas, and walks. The objective may be achieved, but at the expense of increased construction costs, high visual and/or natural impact, user inconvenience, and increased long-term maintenance costs.

The reverse condition is also common. The grid street patterns of many North American cities were plotted with little regard for existing natural and topographic features. The consequences have been the loss of landscape character and uniqueness and the development of a rather homogenized environment. There are exceptions, San Francisco perhaps being the most noteworthy. San Francisco serves as an example of not only a lack of coordination between development and environment but also as an excellent example of contradiction and happenstance. If the city were laid out according to current environmental principles, it may have been denied the charm, drama, and attraction it displays today. The lesson to be learned here is that all principles, even those described in this text, must be viewed as guidelines. A successful designer applies these principles with flexibility and even knows when it is appropriate to bend or break the rules.

Fitting buildings into the landscape is related to the type of structural foundation, the method of construction, and ultimately the form of the architecture employed. There are three general types of structural foundations: slab, continuous wall, and pole (Fig. 4.2). A *slab foundation* forms a horizontal plane with a relatively thin profile, which results in the least amount of grading flexibility. Depending on the length of the building, slopes up to 3% may typically be achieved along the building face. A *continuous wall* foundation forms a line that provides for a moderate degree of flexibility. Depending on the height of the foundation wall, grade changes of a story or more may be achieved. *Pole,* or *pier, construction* provides the greatest amount of grading flexibility and potentially the least amount of grading impact, since there is a minimum

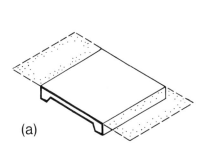 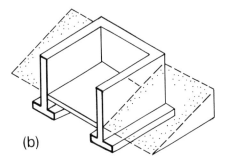 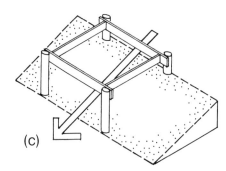

FIGURE 4.2. Foundation Systems. a. Slab. b. Continuous wall. c. Pole.

amount of ground disturbance. This technique uses poles or piers as the primary method of transferring the structural load to the ground. The poles form points rather than lines where they meet the ground, and the building is placed above the landscape, thus allowing drainage to continue uninterrupted. This normally reduces the amount of grading required to redirect storm runoff around structures.

The method of connection between the foundation and the wall system and the material composition of the wall may also create grading constraints. As illustrated in Fig. 4.3a, soil should not be placed directly against wood frame construction. To reduce moisture problems, an 8-in. space is recommended as a minimum between the wood frame and the exterior grade. It should be noted that pressure-preservative-treated lumber may come in contact with soil, but usually this material is used for conditions and purposes other than wood frame construction detailing. Concrete and masonry when properly waterproofed will allow somewhat more flexibility in varying the exterior grade. This is particularly true when the foundation and wall are an integral system as indicated in Fig. 4.3c.

At the risk of considerable oversimplification, architectural design may be categorized as either flat-site buildings or sloping-site buildings with respect to grading. Flat-site buildings are characterized by relatively little change in grade from one side of the building to the opposite side. Sloping-site buildings, on the other hand, attempt to accommodate any existing changes in topography by stepping or terracing the structure with the slope. Each building type is appropriate when used in the proper context. However, an example of an actual case where these simple building-site relationships have been disregarded either through oversight or by the establishment of other functional or aesthetic priorities is illustrated in Fig. 4.4. In this situation, a three-story townhouse complex with garages on the first level was placed on a sloping parcel of land. Entrances to the garages were placed at the rear of the units; this design went contrary to the slope and required additional excavation. This decision may have been based on aesthetic concerns by placing the garage doors behind the units out of view from the street and creating the traditional image of a large lawn in front of the units. Alternatively, if the garage doors had been placed at the front of the units, less excavation would have been necessary, more of the existing vegetation could have been preserved, and a private area could have been provided at the back of each unit at the second-floor level, thus allowing the building to step with the site. This example also illustrates that grading and design are integrated processes and that decision making may involve tradeoffs or compromises.

Often the framework for building placement and development is established by street patterns. There are three basic ways in which street layout is related to topography. The first is to locate the street parallel to the contours. As

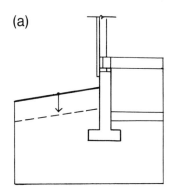 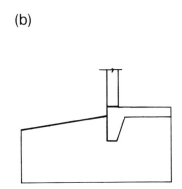 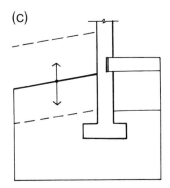

FIGURE 4.3. Exterior Wall Details. a. Standard wood frame construction places wood floor joists on top of a concrete foundation, and the exterior siding material extends below the joist/foundation connection. The grade at the foundation should be at least 8 in. below the siding material. If a continuous wall foundation is used, there is obviously some flexibility downward with the grades. b. Walls are normally constructed directly on slab foundations. Since it is very difficult to waterproof this connection, this condition offers the least flexibility. c. Where the exterior wall forms a continuous plane uninterrupted by floor connections, greater grading flexibility, both up and down, can be achieved.

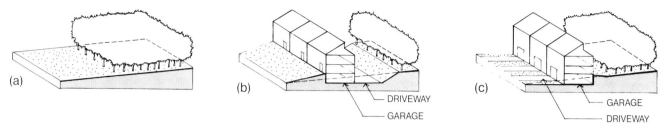

FIGURE 4.4. Relationship of Building to Topography. a. Wooded, sloping site before construction. b. Site as constructed, creating a valley across the slope. c. Alternative site design that reduces the amount of excavation and preserves more of the wooded area.

indicated in Fig. 4.5, buildings placed parallel to the road create obstacles, or dams, for the natural drainage pattern, since there is little pitch in the longitudinal direction. To compensate for this obstruction, the proposed grading must direct the storm water runoff around the buildings. The section that normally results from this arrangement is a terraced condition formed by cutting on the uphill side and filling on the downhill side. The advantage of this layout is that it normally produces easy access between circulation system and building.

The second way is to locate the street perpendicular to the contours. Since the most common practice is to place the long axis of the building parallel to the street, the task of directing storm water around the structure is made somewhat easier by this configuration. However, three potential problems arise from this arrangement. The first is that the gradient of the road may be excessive, since the steepest slope occurs perpendicular to the contours. The second is the awkward relationship that may result at access points from the street, since paths or drives placed perpendicular to the direction of the road must be cross-sloped. Finally, costly grade changes may be required between buildings as illustrated in Fig. 4.6b. Part of this problem may be alleviated by stepping the architecture as shown in Fig. 4.6c.

The third option for street layout is to place the street diagonally across the contours. This arrangement usually provides a more efficient storm drainage design, a better relationship between access and structure, and less steep gradients than those of the perpendicular arrangement. It should also result in the least amount of disturbance to the landscape, since this option should minimize the amount of regrading required.

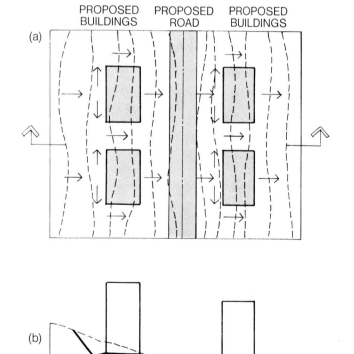

FIGURE 4.5. Road and Long Axis of Buildings Placed Parallel to Contours. a. Plan indicating the proposed drainage pattern. b. Section indicating the typical terrace conditions for this configuration.

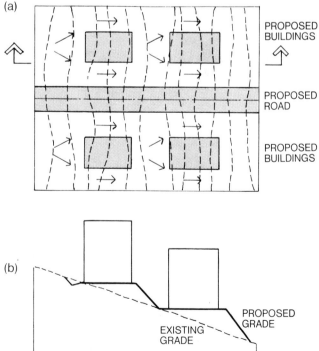

FIGURE 4.6. Road and Long Axis of Buildings Placed Perpendicular to Contours. a. Plan indicating proposed drainage pattern. b. Section indicating stepped terraces.

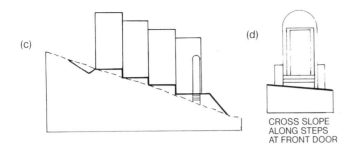

Economics and Maintenance

Economic constraints relate to both initial construction costs and long-term maintenance and operational costs. As an approach to grading design, an optimal solution is one that balances the amount of cut and fill on site. This means that soil does not have to be imported to or exported from the site. If this cannot be achieved, usually the less costly option is to have an excess of cut material. This will be discussed in more detail in Chapter 11.

The types of equipment and labor required during the construction process directly affect cost. Historically, earth moving was accomplished primarily by extensive human effort with the aid of animals and rudimentary machinery. The gardens at Versailles and New York's Central Park are both excellent examples of landscape designs that required large-scale earth moving and grading, executed to a great extent by hand labor. In the twentieth century labor-intensive grading became costly as wages increased and human energy was replaced by powerful, efficient earth-moving equipment. The development of this equipment increased the feasibility for large-scale earth-moving projects, since construction time was reduced and the ratio of the amount of work achievable per dollar spent was greatly increased. Even with the current trends of increased equipment operation costs due to higher energy prices,

FIGURE 4.6. *continued.* c. Section indicating stepped buildings.
d. Elevation illustrating the cross-slope condition at entrances.
e. Height varies along first riser where stair meets steep cross-slope.

More than likely, most development situations will be a composite of the three layout options. The actual grading criteria applied to the grading of any pedestrian or vehicular circulation system depend on many factors. For both systems, such factors as type, volume, and speed of traffic; climatic conditions; and existing topography must be addressed.

Specific activities and land uses also establish limits and constraints on the grading process. Athletic facilities, including playfields and game courts, are excellent examples. Facilities such as football and soccer fields, baseball and softball diamonds, and tennis courts have specific guidelines for layout, orientation, and grading (Fig. 4.8). The preciseness of these guidelines is determined by the anticipated level of play (e.g., international vs. intramural competition or professional baseball vs. Little League) and the governing athletic association. Guidelines and standards extend beyond the world of sports. Highway departments, zoning ordinances, and city, state, and federal agencies have a variety of regulations and standards that must be adhered to in the design and grading process. Table 4.1 lists a number of site conditions and the suggested range of acceptable slopes for each. These guidelines are based on experience and represent common uses under average conditions. They should be applied with flexibility, particularly in unique or atypical situations.

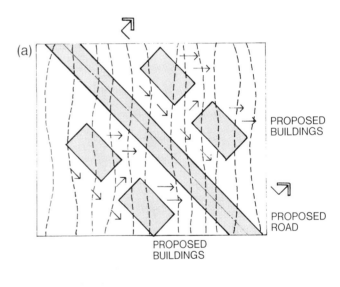

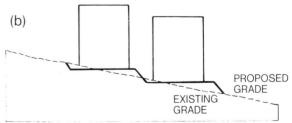

FIGURE 4.7. Road and Long Axis of Buildings Placed Diagonally to Contours. a. Plan indicating proposed drainage pattern. b. Section illustrating a reduced amount of required grade change compared to Fig. 4.5 and Fig. 4.6.

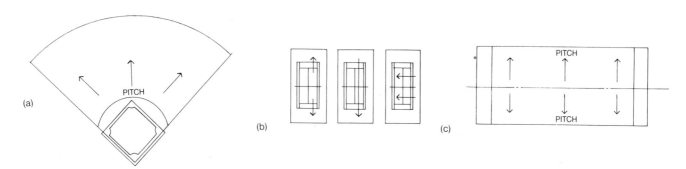

FIGURE 4.8. Grading for Sports Facilities. a. Baseball and softball fields are normally pitched toward the outfield. b. Court sports such as those for tennis and basketball may be pitched from the center to both ends, from one end to the other, or from one side to the other. They may also be diagonally cross-sloped. c. Playfields such as football and soccer fields are crowned at the center and pitched to both sides.

earth moving remains one of the most cost-effective construction processes. Generally, grading designs that use standard construction equipment and a minimum of hand labor in their implementation will be the least costly. Factors influencing the actual method of construction include the configuration of the site and the proposed landforms, available maneuvering and staging space, and preciseness of the grading required.

Grading designs that require elaborate storm drainage systems will be more expensive than designs that conform to natural drainage patterns. In urban areas, it may be difficult to avoid such conditions. In these instances both the grading and storm drainage design must be evaluated for unnecessary intricacy. Simplification, but not at the expense of the overall design concept, will produce the most cost-effective system.

Once a project is completed, it must be maintained. Obviously each design requires a different level of maintenance. Understanding a client's expertise, staffing and budget capabilities, and attitudes toward maintenance at

TABLE 4.1. Grading Standards and Critical Gradients

Type of Use	Extreme Range (%)	Desirable Range (%)
Public Streets	0.5 – 10	1 – 8
Private Roads	0.5 – 20	1 – 12
Service Drives	0.5 – 15[a]	1 – 10[a]
Parking Areas	0.5 – 8	1 – 5
Parking Ramps	up to 20	up to 15
Collector Walks	0.5 – 12	1 – 8
Entrance Walks	0.5 – 8	1 – 4
Pedestrian Ramps	up to 12	up to 8
Stairs	25 – 50	33 – 50
Game Courts	0.5 – 2	0.5 – 1.5
Play Fields	1 – 5[b]	2 – 3[b]
Paved Gutters	0.25 – 100	1 – 50
Grassed Swales	0.5 – 15	2 – 10
Terrace and Sitting Areas	0.5 – 3	1 – 2
Grassed Banks	up to 50[c]	up to 33[d]
Planted Banks	up to 100[c]	up to 50[c]

[a]Preferred gradient at loading/unloading areas is 1 to 3%.
[b]Playfields such as football may be level in the longitudinal direction if they have a 2 to 3% cross slope.
[c]Dependent on soil type.
[d]Maximum slope recommended for power mower. 25% is preferred.

General Comments

1. Minimum slopes should be increased based on drainage capabilities of surfacing materials.
2. Maximum slopes should be decreased based on local climatic conditions, such as ice and snow, and maintenance equipment limitations.
3. All standards should be checked against local Building Codes; Public Work and Highway Departments; and other governing agencies.

the beginning of a project will influence design decisions. Providing a client with a design that cannot be maintained because of impracticality or budget constraints will, in the long run, destroy the project and one's credibility as a design professional. However, there must be a balance between maintenance demands and other aesthetic and functional design criteria.

Among the primary maintenance concerns in grading design are the extent and steepness of sloped areas. Where turf must be mowed, steep slopes will require hand mowers rather than tractor or gang mowers. Improperly maintained steep slopes may also cause soil erosion. Not only is this unsightly, but transported sediment may clog storm drainage structures. This, in turn, creates a safety problem and an additional maintenance concern.

In northern climates, steep slopes along circulation routes may increase the amount of snow removal required and necessitate excessive use of salt or sand to reduce slippery conditions. Also in these climates, where stairs are used extensively and intricate or constricted sidewalk patterns occur, the amount of snow removal done by hand is increased. This slows the clearing process and increases snow removal budgets.

SUMMARY OF CRITICAL CONSTRAINTS

Various constraints and factors influencing grading design decisions have briefly been presented in this chapter. At this point, a few are worth repeating because of their importance.

1. To prevent moisture and structural problems, storm water must be drained away from buildings. This is referred to as *positive drainage*.
2. Grade changes within the drip line of existing trees should be avoided in order to protect the health of the plants.
3. Legally, grades cannot be changed beyond the property lines of the site.
4. The rate of storm runoff leaving the site at any point after construction has been completed should not exceed the preconstruction rate.
5. The proposed grading and landform design should respond to the function and purpose of the activities and uses to be accommodated.

EXERCISES

4.1 As part of their design vocabulary it is important for landscape architects to be able to visualize different percentages of slope and understand how these percentages affect use. One method that aids in developing this skill is to measure percentages of slope for a variety of landscape conditions as they actually appear on the ground. Survey a variety of slopes using the technique illustrated in Fig. 4.9. Try to develop the ability to predict percentages of slope from visual inspection.

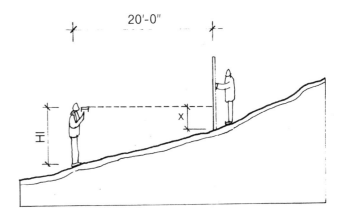

FIGURE 4.9. Field Technique for Quickly Approximating Slopes. A hand level is used to determine the reading on the rod placed 20 ft away. The difference in elevation is obtained by subtracting the rod reading from the height of the instrument, which in this case is the height of the person's eye. The percentage of slope is obtained by multiplying the difference in elevation by 5.

4.2 As discussed in the text, there are relationships among street pattern, building access, and topography. To develop a better understanding of these relationships, examine your surrounding community. Look at how topography may have influenced street patterns and how the means of buildings access change as topographic conditions change.

4.3 Grading around buildings is related to the type of structural foundation, method of construction, and building form. Examine a variety of building conditions to see how grading is handled with respect to building entrances, construction materials, and types of foundation. Look at how the building has been adapted to, or imposed on, the topography (i.e., flat-site vs. sloping-site building).

4.4 During a rainstorm, examine how grading directs the flow of surface runoff. Also examine how surface conditions (paved, unpaved, roofs, etc.) influence the rate and amount of runoff. In addition, look for drainage problems. Analyze why they occur and how improvements might be made.

5
Grading Design and Process

GRADING DESIGN

It is extremely important to realize that grading is one of the primary design tools available to the landscape architect. Every site design project requires some change in grade. How these grade changes are integrated into the overall design concept will influence the success of the project both functionally and visually. The necessity for grade changes goes beyond the constraints discussed in Chapter 4 to include aesthetic, perceptual, and spatial considerations.

Aesthetics

The visual form of grading may be broadly categorized into three types. The selection of a particular type is appropriate within a given landscape or design context, but it is possible to combine types within the same project. The three categories are geomorphic, architectonic, and naturalistic (Fig. 5.1).

(b)

(a)

(c)

FIGURE 5.1. Visual Form of Grading. a. Geomorphic: The geologic formation of Manhattan Island is expressed in the grading design of New York's Central Park. b. Architectonic: The sloping grassed plane is used to reinforce the architectural edge of this space. c. Naturalistic: In this large urban park the land was manipulated to create a meadow within a valleylike space.

Geomorphic

The proposed grading blends ecologically and visually with the character of the existing natural landscape. It reflects the geologic forces and natural patterns that shape the landscape by repeating similar landforms and the physiographical structure. Generally, the intent of this category is to minimize the amount of regrading necessary in order to preserve the existing landscape character.

Architectonic

The proposed grading creates uniform slopes and forms, which usually are crisply defined geometric shapes. The line along which planes intersect is clearly articulated, rather than softened by rounded edges. This type of grading is appropriate where the overall impact is human-dominated or where a strong contrast between the built and natural landscape is desired.

Naturalistic

This last category is perhaps the most common type of grading, particularly in suburban and rural settings. It is a stylized approach in which abstract (or organic) landforms are used to represent or imitate the natural landscape.

Perception

Slope

The perception of slope is influenced by the texture of the surface material and the relationship to surrounding grades. The coarser the texture, the less noticeable the slope. For example, the slope of smooth pavement, such as troweled concrete, is more noticeable than that of coarse pavement such as cobblestone. Generally, slopes of 2% or greater on pavements can easily be perceived. However, horizontal reference lines, such as brick coursing or the top of a wall, increase the awareness of slope even in unpaved situations.

The relationship of one slope to another will also influence perception of steepness. For example, when traveling along a walk with an 8% slope, which then changes to a 4% slope, the 4% slope will visually appear to be less than half the original slope.

Topography, landform, and change in grade break the landscape into comprehensible units, which establish a sense of scale and sequence. The manner in which these grade changes occur affects spatial and visual perception and image of a place.

Elevation Change

Being at a higher elevation relative to the surrounding landscape is potentially dramatic for a variety of reasons. First, a rise in grade may provide a feeling of expansiveness by extending views and overall field of vision. Being at a higher elevation may also provide a sense of superiority, which may contribute to a feeling of control or dominance of a place. In addition, an upward change in elevation can provide an opportunity to contrast or exaggerate the steepness or flatness of the surrounding landscape. The abruptness of the elevation change will affect how space is perceived. The more gradual the ascent the more subtle the experience. The steeper the grade the greater the sense of enclosure at the lower elevations, while the opportunity for drama and excitement is increased at the higher elevations.

Convex and Concave Slopes

Generally a plane is visually less pleasing than a convex or concave landform, although this depends on scale and contrast (Fig. 5.2). In comparing the two rounded forms, the concave appears more graceful from the downhill side, since it exhibits an uplifting quality. From the downhill side both forms foreshorten the view, with the foreshortening much more abrupt with the convex slope. From the uphill side of a convex slope the sense of height is accented. Also, the sense of distance appears compressed, since the middle ground is foreshortened. The following quote from Arthur Raistrick's *The Pennine Dales*, which appeared in *The Sense of Place* by Fritz Steele, illustrates the influence landform can have on experience:

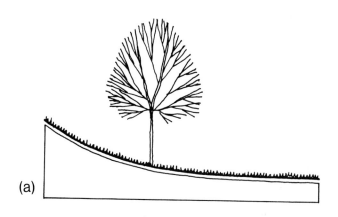

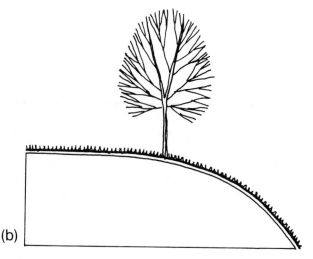

FIGURE 5.2. Rounded Slopes. a. Concave. b. Convex.

The most impressive approach to a view of one of the dales is to come upon it from the high moors—what the dalesfolk call, so expressively, from off the "tops." One has spent the day, perhaps, up in this world of heather, with grouse or curlew providing a commentary to one's movement, and with wide views of moorland cut by faint runnels and gullies, many of which are, in fact, the gaps of the dale lip seen in foreshortened perspective. The high ground begins to decline and one may come to the edge of the heather and peat and enter a world of benty grass and occasional stream heads. Then comes the moment when one looks "over the edge"—the convexity of the hill has reached the point where one can look back up the gentler slope of moorland, or forward down what often appears to be an almost precipitous slope into the valley.

Enhancement

From analyzing the existing topography and landscape character, proposed landforms, grade changes, and design elements may be constructed or placed to emphasize, negate, or have little impact on the visual structure of the landscape. The basic considerations when proposing design alternatives are whether they will enhance, complement, contrast, or conflict with that particular landscape context (Fig. 5.3).

Spatial Considerations

Proposed grade changes may perform a variety of spatial functions. The appropriateness of the application of these functions is determined by a careful analysis of the potential of the site and the demands of the design program. Several grading design applications are discussed and illustrated in this section.

Enclosure

Enclosure may be used to perform several tasks, including containment, protection, privacy, and screening. Seclusion, intimacy, and privacy may be achieved through the use of containment, as in the section of the sitting area illustrated in Fig. 5.4. Screening is a form of visual containment, since it terminates sight lines and eliminates undesirable views (Fig. 5.5).

Enclosure, possibly in the form of a berm, as illustrated in Fig. 5.6, also may be used to provide safety and protection, such as restraining children from running into the

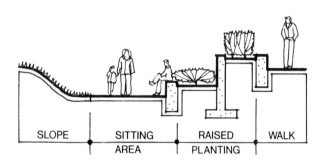

FIGURE 5.4. Level Changes for Privacy. Raised planting separates and visually screens the sunken sitting area from the sidewalk. The slope is used to add to the spatial enclosure of the sitting area.

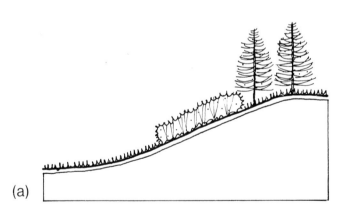

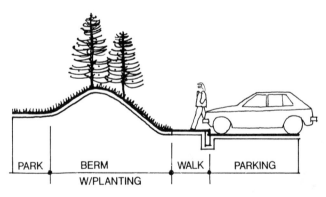

FIGURE 5.5. Visual Screen. Topography, particularly in conjunction with planting, can be used to screen or block undesired views. In the illustration, a planted berm is used to screen the view of a parking lot from a park area.

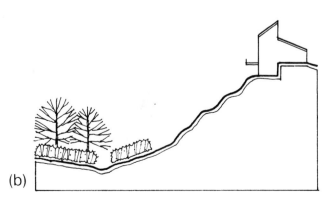

FIGURE 5.3. Enhancing Topography with Design Elements.
a. Planting. b. Architecture.

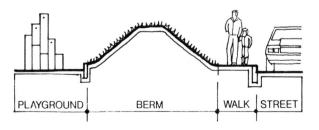

FIGURE 5.6. In the section a berm is used to separate a playground from a street.

street from a playground or serving as a backstop for various types of athletic activities. However, this type of application should be used with caution for two reasons: First, it may inadvertently promote careless recreational uses, and, second, the enclosure reduces visibility into the area, creating a potentially unsafe condition. It should be noted that properly designed and placed landforms can be an excellent outlet for creative play.

Berms are vegetated or paved embankments, somewhat dikelike in appearance, commonly used by landscape architects for enclosure and separation purposes. However, the use of these devices must be carefully evaluated, since there are many examples where the scale and proportion of berms have been insignificant or inappropriate with regard to the surrounding context.

Enclosure may also provide protection from climatic elements (Fig. 5.7). Properly placed landforms can control drifting snow and significantly reduce the impact of wind on structures and even on large areas such as playfields and parking lots.

Separation

A very basic application of grading is the separation of activities to reduce potential conflicts, such as separating auto traffic from pedestrians and bicyclists, bicyclists from pedestrians, or sitting areas from walkways. In the design of New York's Central Park, Olmsted and Vaux showed great vision in using grade changes and overpasses to separate the traffic on the transverse roads from the park. Even a change in grade of only a few inches may sufficiently define the territory in which an activity may occur. Separation may be accomplished by a variety of techniques, two of which are illustrated in Fig. 5.8.

Channeling

Landforms may be used to direct, funnel, or channel auto and pedestrian circulation. They may also be used to direct and control viewing angles and vistas as well as wind and cold air drainage. An amphitheater is a special use of

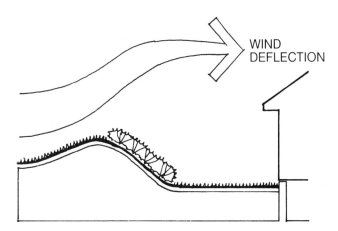

FIGURE 5.7. Microclimate Modification. Topography can be used to channel or deflect winds, capture solar radiation, and create cold or warm pockets.

landform both to focus attention and to enclose space. See Fig. 5.9.

It must be realized that in all the applications listed, the functions of landform and grade changes are reinforced and strengthened by the use of plantings and structural elements such as walls and fences.

GRADE CHANGE DEVICES

Accommodating changes in elevation in outdoor environments is one of the basic functions of a landscape architect. There are many devices that can accomplish these changes, including stairs, ramps, perrons, walls, slopes, and terraces. Concern for the experiential, functional, and visual manner in which these grade transitions are executed should result in good landscape design.

Stairs

Use of stairs is the most common technique to accommodate pedestrian circulation where an abrupt change in

(a) (b)

FIGURE 5.8. Separation of Activities. a. Along a riverfront promenade a grade change, which incorporates seating, is used to separate the pedestrian walk from the bicycle and service lane. b. An underpass is used to separate pedestrian and vehicular circulation in New York's Central Park.

grade is necessary or desired. The width of a stair depends on the design intent but should be a minimum of 3 ft. A desirable minimum width to allow two people to pass comfortably is 4 ft.

The proportion of riser height to tread width is critical to the ease and comfort of the user. A standard rule of thumb that has evolved for *exterior* stairs is that two risers plus one tread should equal 24 to 26 in. Two commonly used ratios are 6-in. riser with 12-in. tread and 5-in. riser with 15-in. tread. Preferably, there should be a minimum of 3 risers and a maximum of 10 to 12 risers for a set of stairs. The minimum number is suggested to make the stair more visible to prevent tripping. Where the number of risers exceeds 12, an intermediate landing is recommended to reduce the apparent scale of the stair and provide a comfortable resting point. Again, both of these guidelines must be applied with flexibility. Handrails are normally required on stairs with five or more risers, although this point must be checked against local building code requirements. Stair treads should pitch 1/8 in./ft in the downhill direction to ensure proper drainage. Low

FIGURE 5.9. Amphitheaters. There are many good examples where topography has been used effectively to create a theater setting, two of which are presented here. a. A Roman amphitheater built in Caesarea, Israel, about 25 B.C. The theater has been reconstructed and is in use today. b. An amphitheater constructed at the site of the 1972 Olympics in Munich. Note how the edge of the theater has been blended into the surrounding earthform.

curbs, referred to as *cheek walls,* are commonly used along the edges of stairs for safety and maintenance purposes. See Fig. 5.10.

The configuration of contour lines where they cross stairs is illustrated in Fig. 5.11. However, contour lines are drawn only to the outside edge of the cheek walls on grading plans, and the grades at the top and bottom of the stair are indicated by spot elevations. It is not necessary to indicate the spot elevation for each step, since this information is provided by dimensions on construction detail drawings.

Ramps

Ramps are simply inclined sidewalks or driveways, usually with uniform slopes. Typical slopes range from 5% to 8% for pedestrian use and can be as great as 15% where handicapped access is not a constraint. Again, 4 ft is a recommended minimum width to allow people to pass. Handrails, particularly for handicapped use, are normally required on ramps with slopes of 5% and greater. Where ramps exceed 30 ft in length, an intermediate landing is usually required. Ramps may be graded to slope perpendicular to the direction of travel (or cross-pitched) for drainage purposes. Contours are drawn across ramps on grading plans, and spot elevations are used to indicate top and bottom of ramp.

Stairs and ramps may be used in combination. The arrangement may be sets of stairs alternating with ramps, as illustrated in Fig. 5.13a, or lengthened and ramped treads, as in Fig. 5.13b. Many examples of the latter arrangement, most of which are uncomfortable to use, can be found. Not only are the height of the risers and the distance between risers important in this situation, but so is the percentage of slope between risers, since the slope shortens a person's stride in the uphill direction and lengthens it in the downhill direction.

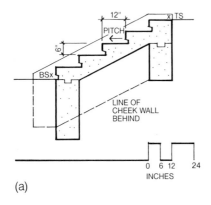

(a)

(b)

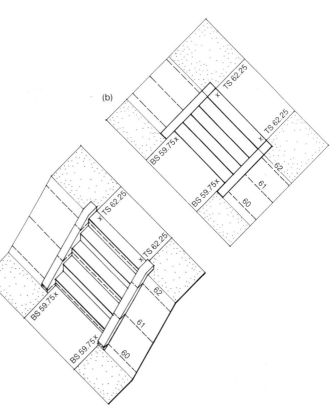

FIGURE 5.10. Stairs.
a. Typical stair section.
b. Stair spiraling up a steep slope.
Both the stair and the path are pitched
to a gutter on the uphill side,
which also intercepts the runoff from the slope.

FIGURE 5.11. Grading at Stairs. a. Plan oblique illustrates how contour lines follow along the face of the stair risers. b. On construction drawings the contour lines are drawn only to the edge of the stairs or cheek walls. Spot elevations are given at the top and bottom of the stairs and *not* for each step.

Retaining Walls

Retaining walls allow for the greatest change in elevation in the shortest horizontal distance. They are also the most expensive method for accommodating grade changes but may be necessary where space is at a premium, such as at urban or small sites or where the use of walls is an integral part of the design concept.

Walls must be structurally designed to support the weight of the earth being retained; however, structural calculations and sizing of retaining walls are beyond the scope of this text. Proper drainage of retaining walls is critical to their stability. There are two primary issues related to drainage. The first is the buildup of water pressure behind the wall, which, if not alleviated, will cause the wall to slide or overturn. To prevent buildup, drain holes, referred to as *weepholes,* are constructed near the base of the wall or a lateral drain pipe is placed behind the wall to collect and dispose of excess groundwater. The second issue is the saturation of the soil under the wall, which can cause overturning due to reduced bearing capacity. This is prevented by intercepting storm runoff on the uphill side of the wall and can be accomplished through the use of swales or drains. Preventing runoff from flowing over the top of the wall also reduces potential staining of the face of the wall.

Figure 5.15 illustrates the drawing of contour lines where they intersect a vertical surface such as a retaining wall. On a grading plan, the contour lines are drawn to the face of the wall but do not need to be drawn along the face, since they would not be visible. Top and bottom of wall spot elevations are typically indicated at the ends and corners of walls.

Several ways in which sloping ground and retaining walls may interface are illustrated in Fig. 5.16. A slightly exaggerated pitch at the base of retaining walls to ensure positive drainage is indicated by the contour lines where they abut the wall. The condition illustrated in Fig. 5.16e is unacceptable because of erosion and maintenance problems, unless the wall is quite thick.

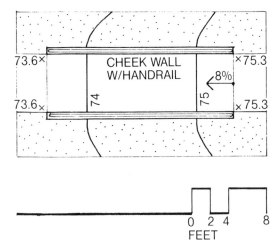

FIGURE 5.12. Grading at Ramps. Spot elevations are indicated at the top and bottom of a ramp, and contour lines are drawn across the ramp. An arrow, normally pointing downhill, is shown with the slope of the ramp. It should be noted that the slope on ramps to be used by the handicapped cannot be more than 8%; this is approximately 1-in. vertical change for each 1 ft of horizontal distance. Thus the length of a ramp sloped 8% will be 12 ft for each 1 ft of elevation change.

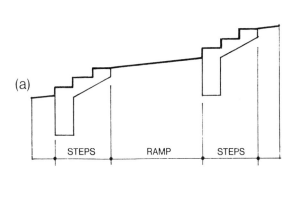

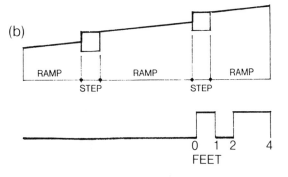

FIGURE 5.13. Combination Ramp Stairs. a. Sets of stairs connected by a short ramp. b. Alternating arrangement of ramp and step. c. Photograph illustrates the condition in part b. Again, the walk has been pitched toward a gutter on the uphill side.

52 Grading Design and Process

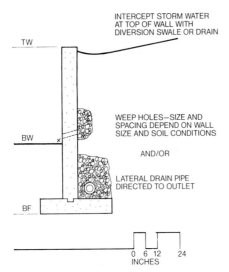

FIGURE 5.14. Retaining Wall Section. Depending on the conditions, weep holes, a lateral drain, or both, may be required to prevent a buildup of water pressure behind a wall.

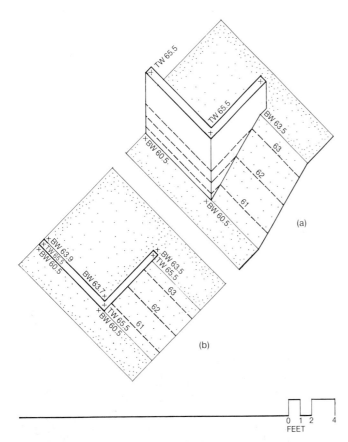

FIGURE 5.15. Contour Lines at Retaining Walls. a. Similar to curbs and stairs, the contour lines again follow along the face of the structure. b. On construction drawings contour lines are drawn to the edges of the wall. Since these lines are superimposed on the face of the wall, they are not seen. Spot elevations are indicated at appropriate places.

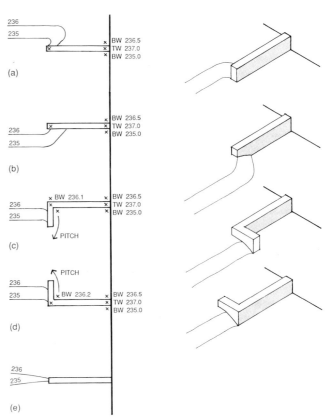

FIGURE 5.16. Grading Options at Retaining Walls.
a. A slope is created along the uphill face of the wall, thus making the end of the wall more visually apparent.
b. A slope is created along the downhill face, somewhat reducing the scale of the wall visually.
c. The wall is shaped in the form of an L, with the L pointing in the downhill direction. A niche is formed on the downhill side, and again the end of the wall protrudes as in part a.
d. Again, the L form is used, but now pointing in the uphill direction. As a result, the outside corner of the L becomes more apparent.
e. Except for wide walls, this condition is unacceptable because of erosion, maintenance, and safety problems. In the options in parts c and d, the grading must be handled to prevent the trapping of storm water at the inside corner of the L.

SLOPES

Slopes are the least costly technique for changing grade but require more space than retaining walls. To be visually significant, slopes should not be flatter than 5 to 1. Generally, planted slopes should not exceed 2 to 1, but paved slopes may be 1 to 1 or steeper. Mowed lawn areas should not exceed 3 to 1, although 4 to 1 is the preferred maximum. The use of slopes and the selection of a desired gradient are based on the design intent, soil conditions, susceptibility to erosion, and type of surface cover. All slopes must be stabilized by vegetative or mechanical means to reduce erosion potential.

Terraces

Terraces provide a series of relatively flat intermediate levels to accommodate a change in grade. The reasons for terracing may be visual, functional, or structural. Terraces may be created by the use of slopes or walls or a combination of both (Fig. 5.17). Where both the slope and the level portion, known as a *bench*, are relatively small in area and grade change, it is possible to pitch the bench in the downhill direction. However, where grade changes and area of slope are considerable, as in a highway cut or fill, the bench must be pitched back from the slope to reduce erosion (Fig. 5.18). The storm runoff from the bench must be properly disposed of to prevent saturation of the toe of the uphill slope, which could cause the slope to slump.

GRADING PROCESS

The grading procedure presented in this text is best described as a *controlled intuitive process*. Unfortunately, there is no precise sequence to follow in order to arrive at a correct grading solution, since every grading problem is different and in most cases there is more than one appropriate solution to a particular problem. This lack of precision is often a source of frustration to students. Frustration may be reduced through continued practice and by a generalized three-phased approach. These phases include (1) an inventory and analysis of the site and development program, (2) design development, and (3) design implementation.

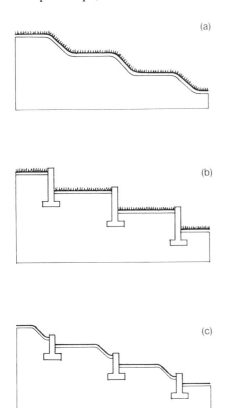

FIGURE 5.17. Terrace Sections. a. Terraces created by slopes. b. Terraces created by retaining walls. c. Terraces created by combining slopes and retaining walls. d. An example of part c.

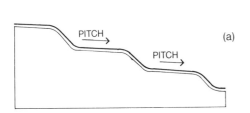

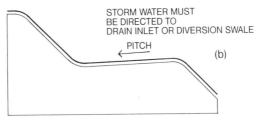

FIGURE 5.18. Drainage for Slopes and Terraces. a. For relatively small terraces the bench may be pitched in the downhill direction so that storm runoff flows across the slopes. b. For large terraces, steep slopes, or easily erodible soils the bench should be pitched back from the top of the slope. It is important, however, to prevent saturation of the bench by properly disposing of storm runoff.

Phase 1: Inventory and Analysis

The initial phase of any landscape architectural design project is to inventory the existing physical and cultural conditions of the site and its surroundings. Both the inventory and the development program (i.e., what is to be placed on the site) are analyzed and evaluated to identify conflicts, constraints, and opportunities that will guide the design development and the proposed grading scheme. This phase relates directly to the discussion in Chapter 4 and the preceding sections of this chapter.

Phase 2: Design Development

During phase 2 the observations and evaluations of the first phase are synthesized into a design concept that may generate one or more design solutions. As a result of the compartmentalization of the educational system, which typically separates courses in grading technology from courses in design, this phase is often overlooked by students. From experience and observation, many students wrongly conclude that, because grading is taught as an engineering subject, design is of secondary importance. The reverse situation is commonly found in the design studio.

With regard to grading and design, three points must be emphasized: First, grading and site design are two highly related and dependent processes. To achieve an appropriate as well as successful final product, both must be integrated in a holistic manner at the outset of the project. Second, before manipulating contours on a grading plan it is important to have a clear understanding of the form of the desired final product. Without this knowledge the manipulation of contours is an aimless and futile act. To reinforce this point, any appropriate three-dimensional form can be expressed by contours on a grading plan. However, without a preconception of what that form should be, it can never be attained. Finally, a change in grade must be purposeful, whether for functional or aesthetic reasons, and not arbitrary. The intent to change a grade 2 in. is no less important than the intent to change a grade 20 ft.

Phase 3: Design Implementation

At this point the design solution is translated into a workable and buildable scheme through the development of construction drawings and specifications. Although not always applicable or necessary, four steps may be followed to make the task of developing accurate, concise, and correct grading plans easier. Based on a process developed by David Young and Donald Leslie at Pennsylvania State University, these steps include (1) development of section criteria, (2) application of section criteria, (3) development of slope diagrams, and (4) evaluation of slope diagrams.

Since grading is the change upward or downward of the ground surface, plans alone are normally inadequate for studying vertical relationships properly. Therefore, sections should be drawn at critical points and section criteria developed to establish relative vertical relationships and approximate slopes, and select appropriate grade change devices such as walls, slopes, and steps. Once the section criteria have been developed, they may be applied to the plan to establish key spot elevations.

The purpose of slope diagrams is to outline the extent of areas to be graded to relatively the same slopes, indicate the direction of proposed slopes, establish drainage patterns by outlining proposed drainage areas and locating proposed high points and collection points, and, finally, establish spot elevations at critical points such as building corners and entrances, top and bottom of stairs and walls, and high points of swales. For evaluation purposes, a checklist may be established to determine whether the slope diagrams maintain the integrity of the overall design concept; respond appropriately to identified constraints; provide positive, efficient, and ecologically sensitive drainage; and maintain a relative balance between cut and fill, if possible. Slope diagrams should be revised where problems or conflicts are identified.

It should be noted that in the steps outlined no contour lines have been manipulated or drawn. Only areas to be regraded, approximate slopes, and critical spot elevations have been determined. These now provide a controlled framework within which the proposed contour lines may be located.

It is important to realize that very rarely is a grading plan completely correct or appropriate on the first try. The development of a grading plan is to some degree a trial and error process often requiring numerous adjustments. These adjustments are usually minor; however, in some instances the entire grading approach may need to be re-examined. Finally, although the overall design concept is established during the design development phase, many detailed design decisions made at the implementation phase will contribute to the ultimate success of a project.

Example 5.1

The following example is presented to illustrate a typical sequence of events associated with a grading design project. In this project a small, L-shaped office building with a slab foundation is to be constructed on a suburban corner lot. The development program requires a seven-car parking area with potential for expansion, an entrance court, and handicapped access. The site is to be designed and the grading plan developed.

Site Analysis

As can be seen in Fig. 5.19, there are no major natural constraints on the site. The high point located in the northeast corner is at elevation 44.8 ft, while the low point, elevation 35.8 ft, is located on the property line in the southwest corner. The steepest slope is approximately 10%; the average slope across the site is approximately 6%. The existing surface drainage pattern is primarily sheet flow, with approximately half directed toward Third Street and half toward Oak Street. The most significant feature on the site is the grove of mature red oak trees.

The adjacent land uses include a bank on the north and a single-family residence on the east. The residence is screened from the property by a hemlock hedge. The heaviest vehicular traffic is along Oak Street. Most pedestrians will approach the site from the southwest.

Design Development

As shown in Fig. 5.20, both the building and parking are located as far from the oak trees as possible to ensure their preservation. The building is placed with the open part of the L toward the street corner in order to orient the building entrance toward the primary pedestrian approach and to maximize the amount of glazing facing south to increase energy efficiency as this site is located in a cold climate region. The entrance court is developed within the open space of the L. The parking area is located so as to maximize the distance between the driveway entrance and the street intersection as well as to take advantage of the visual screen already provided by the adjacent hedge. The parking entrance is not located on Oak Street, in order to preserve the existing oak grove and to prevent possible conflicts with the bank traffic.

Design Implementation

In developing the grading plan the four major elements of

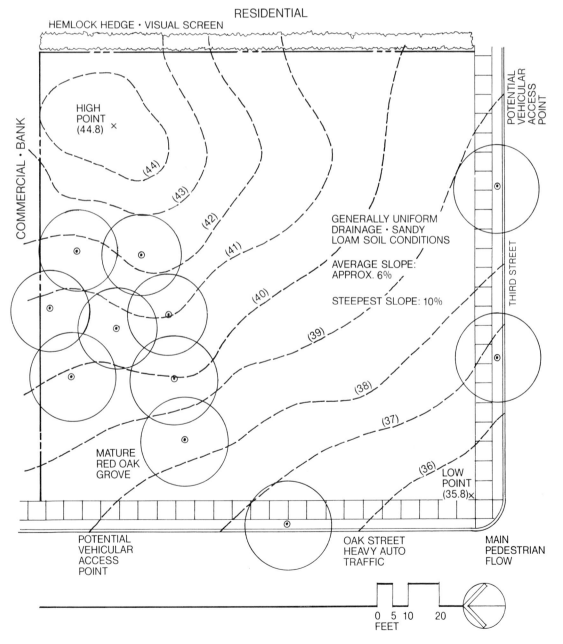

FIGURE 5.19. Site Analysis for Example 5.1.

56   Grading Design and Process

the design, the parking area, building, entrance court, and oak grove, are analyzed. Where appropriate, section criteria and grading diagrams are developed for each component and alternative solutions are presented and explored.

Parking Area. The existing grade slopes downward toward Third Street along the longitudinal axis of the parking area. To be consistent, the proposed grading for the parking area should slope in the same direction. There are four basic approaches to grading the parking area, as illustrated by the sections and slope diagrams in Fig. 5.21. The first is to cross-pitch the pavement toward the building and collect the storm water runoff in the southwest corner of the lot. The second is to cross-pitch the pavement away from the building and collect the runoff in the southeast area of the lot. The third combines the first two options, resulting in a valley down the center of the parking area. The fourth alternative again combines the first two options to create a ridge down the center of the parking area. Runoff is collected in both the southeast and southwest corners of the lot.

Generally, cross-sloping small parking areas in one direction is the most efficient and visually least disturbing method of grading. However, in working through the four options for this problem, it was determined that pitching toward the center was the best alternative. Cross-pitching the parking area toward the building placed the catch basin too close to the building and in an inconvenient location for pedestrians, particularly if it clogged. Cross-pitching

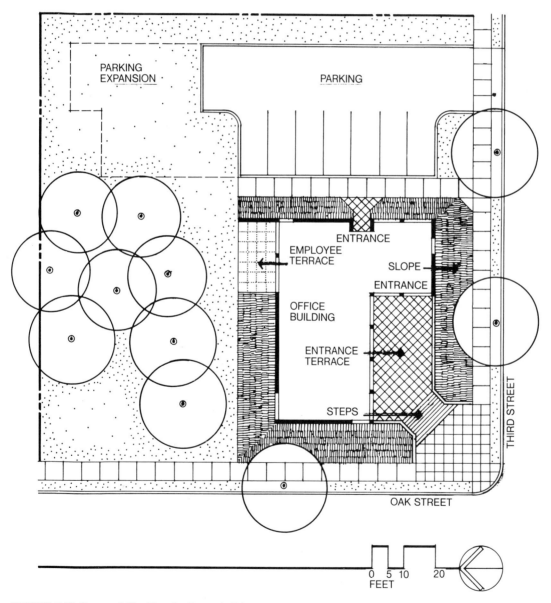

FIGURE 5.20. Proposed Site Plan for Example 5.1.

Grading Process 57

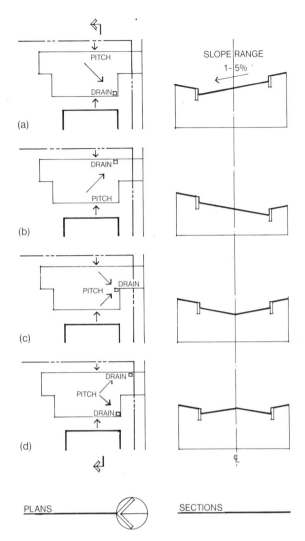

FIGURE 5.21. Alternative Slope Diagrams and Sections for Parking Area. a. Cross-pitched toward building. b. Cross-pitched away from building. c. Pitched to center, creating a valley. d. Pitched to edges, creating a ridge.

vation must be evaluated. A basic slope diagram for the building is shown in Fig. 5.22.

In examining the relationship of the building to the existing grades, a finished floor elevation of approximately 38.0 ft would seem appropriate, since the 38-ft contour line roughly bisects the building. This is a typical strategy, since it strikes a balance in the grade change from one end of the building to the other and potentially produces a more balanced cut and fill condition. However, when examining the relationship of the east entrance to the parking area, there are two basic alternatives. The first is to create a positive drainage condition; thus the elevation at the edge of the building must be higher than the elevation at the top of the curb at the edge of the parking lot as shown in Fig. 5.23a. If the finished floor elevation is established at 38.0 ft, the elevation of the parking lot adjacent to the east entry must be lower than 38.0 ft. This results in approximately 2 ft of cut in the parking lot. A second alternative would be to construct the walk and edge of the parking area at an elevation above the finished floor elevation as illustrated in Fig. 5.23b. This results in more costly construction and may cause drainage problems, since a low area is created directly adjacent to the building.

In the third and final analysis, perhaps the finished floor elevation (FFE) should be raised above 38 ft, thus ensuring good drainage, reducing cut, and lowering construction costs. In raising the elevation of the building, however, most of the foundation slab will be placed on fill. Purchasing and properly placing the fill material can add to the cost of the project. Also, if the fill material is not compacted properly, future soil settlement may cause cracking and structural problems within the building. As can be seen from these three conditions, a number of factors must be explored and assessed in making grading decisions. In this case, since the scale of the building is

away from the building resulted in more cut in the northeast corner of the lot. The ridge alternative did not result in significant grading advantages and would be more costly because of the additional storm drainage structure. Other factors that influenced this decision are the accessibility for the handicapped at this end of the building and the limited change in grade that can occur along the face of the building caused by the slab construction. The controls for selecting the proposed elevations and gradients within the parking area are based on the preceding factors in conjunction with the existing grades along the east property line and at the entrance to the parking area.

Building. Several concerns must be addressed in establishing a finished floor elevation (FFE) for the building. First, positive drainage away from the building must be maintained on all sides. A second concern is the elevation of the east entrance and its relationship to the proposed parking lot grades. The elevation of the southwest entrance and its relationship to existing sidewalk grades must also be considered. In addition, the proposed finished floor ele-

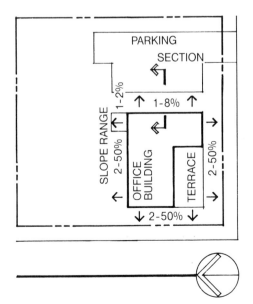

FIGURE 5.22. Slope Diagram for Building. The primary objective is to maintain positive drainage away from the building. Except for the east side, where handicapped access must be maintained, there are very few slope restrictions.

58 Grading Design and Process

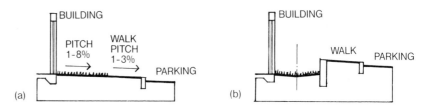

FIGURE 5.23. Section at East Face of Building. a. Floor elevation set higher than parking lot elevation. b. Floor elevation set lower than parking lot elevation.

small, it is felt that the third alternative (i.e, raising the FFE) represents the best option and could be properly constructed.

Entrance Court. There are several ways in which the entrance court could be designed, each directly related to grading decisions. Three schematic plans and accompanying sections are illustrated in Fig. 5.24. By establishing the finished floor elevation above elevation 38.0 ft, it is necessary to provide stairs at the southwest entrance. Also, along the south and west faces of the building slopes are needed, since the floor elevation is above the existing grades of the sidewalk. The three solutions illustrated respond to these conditions. It should be realized that these solutions were studied during the design development phase in order to reach a design decision. Again there is

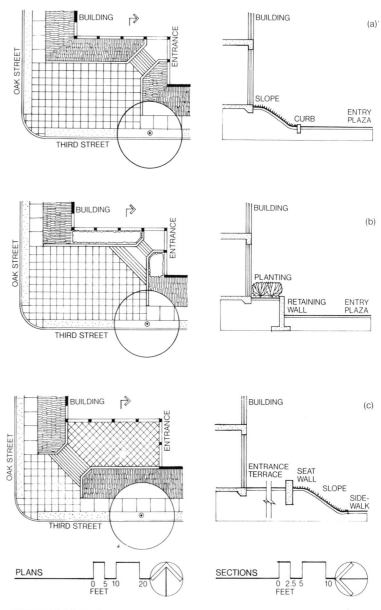

FIGURE 5.24. Design Alternatives for Entrance Terrace.

not always a distinct separation between the types of work to be performed in each of the phases. The third option (Fig. 5.24c) was selected for several reasons. The entrance courts in Fig. 5.24a and Fig. 5.24b were considered too large in relation to the size of the building. The option in Fig. 5.24c provides a transitional space between the street corner and the building entrance, and, by continuing the slopes along the edges of the entrance court, the relationship of the building to the street corner is more clearly defined. The other two solutions are also correct and could be implemented if different design objectives were desired.

Oak Grove. In order to preserve the oak grove, there should be as little regrading and soil disturbance as possible within the drip line of the trees. Minimizing grading restricts the extent to which the existing contour lines may be altered on the north side of the building. As a result, the intent is to return to the existing grade in as short a distance as possible beyond the building. To be consistent with the treatment of the south and west sides, a slope is used along this edge of the building.

Synthesis

The rationale and criteria presented in the previous sections are synthesized into a final grading plan. Figure 5.25 illustrates the placement of critical spot elevations and proposed gradients; Fig. 5.26 illustrates the final grading plan. From this drawing a grading contractor should be able to execute this portion of the project. Again it must be emphasized that this is *a* solution, not *the* solution, to this problem.

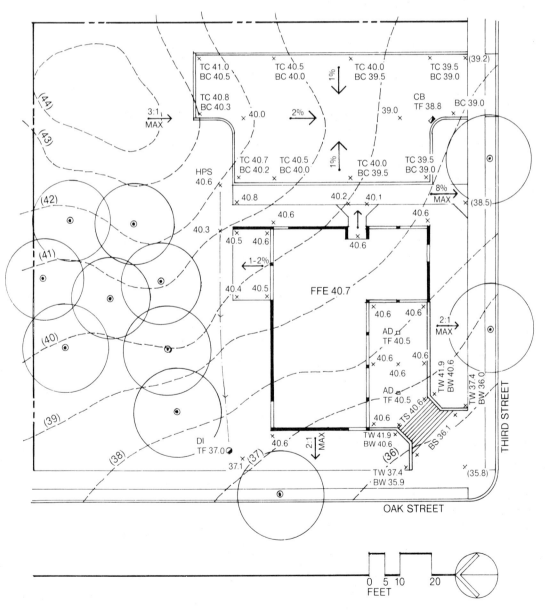

FIGURE 5.25. Critical Spot Elevations and Gradients.

GRADING PLAN GRAPHICS

There are two basic types of grading plan. The first is the conceptual grading plan, which communicates the design intent but is not usually an accurate or engineering representation of the ground form (Fig. 5.20). The audience for this plan is normally the client, who may be an individual, architect, or public agency, and its purpose is to make the proposed concept easily understandable. The second is the grading plan executed as part of a set of construction documents (Fig. 5.26). The purpose of this drawing is to interpret the design intent accurately and communicate this information to the grading contractor effectively. The plan, in conjunction with the technical specifications, must provide complete instructions as to the nature and scope of work to be performed as well as a solid basis for estimating the cost involved. The success

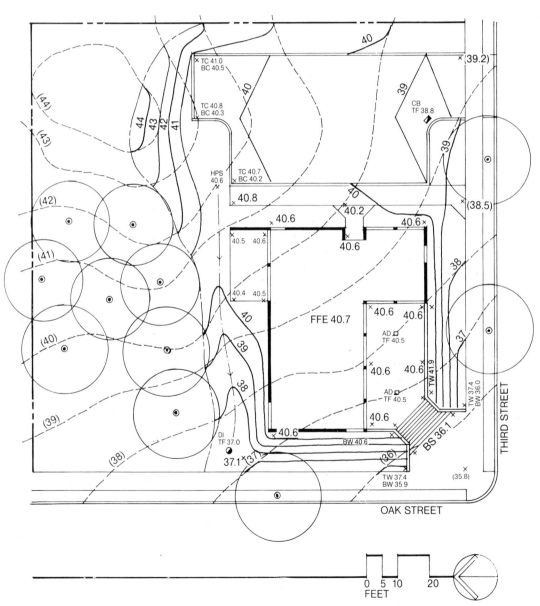

FIGURE 5.26. Final Grading Plan.

of a project depends on the *accuracy, completeness,* and *clarity* of the construction drawings.

Construction Grading Plan

The grading plan should show all existing and proposed features of the site. This includes all buildings; structures, such as walls, walks, steps, and roads; utilities, such as water, sewer, storm drainage, and electrical lines; utility structures, such as manholes, meter pits, and junction boxes; and all other underground structures, such as vaults, septic systems, and fuel storage tanks. Proposed features are normally drafted as solid lines and existing features are shown as dashed lines or are photographically screened to appear lighter. In addition, of course, both existing and proposed contour lines and spot elevations are shown.

Spot elevations are used to supplement contours in the following situations:

1. To indicate variations from the normal slope or gradient between contour lines.
2. To indicate elevations of intersecting planes and lines, such as corners of buildings, terraces, and walks.
3. To indicate elevations at top and bottom of vertical elements, such as walls, steps, and curbs.
4. To indicate floor and entrance elevations.
5. To indicate elevations of high and low points.
6. To indicate top of frame (rim) elevations and inverts for utility systems.

With the preceding list in mind, examine the use of spot elevations in Fig. 5.26. In some cases only spot elevations are shown on highway plans and contour lines are omitted.

The following elements should appear on all grading plans:

1. *Written and/or graphic scale:* The scale at which a grading plan is drawn depends on the scope of the project and the nature of the available topographic data. Scales for site plans usually range from 1/8 in. = 1 ft to 1 in. = 40 ft.
2. *North arrow:* A north arrow is provided for orientation purposes. It should be indicated whether this is assumed, magnetic, or true north.
3. *Notes:* Notes include general or explanatory information as well as descriptions of any unique conditions of the plan. All plans should contain a note describing the source from which the existing conditions were taken as well as bench marks and reference datum.
4. *Legend:* All symbols and abbreviations used on the drawing should be identified in a legend. Examples of typical symbols and abbreviations are presented in Fig. 5.27. It should be pointed out that convention varies in different regions of the country and even between design offices within the same region. This fact only reinforces the need for a legend. The conventions used in this text include dashed lines for existing contours, solid lines for proposed contours, parenthetic labels for both existing contours and existing spot elevations, labeling contours on the *uphill* side, and use of a wider line for every fifth or tenth contour line. Generally, contour lines are drawn freehand for unpaved areas and are drafted for paved areas.
5. *Title block:* Information such as project name, location of project, name of client, name of design firm, drawing title, drawing number, scale, and date should be arranged in an orderly and easily readable manner, normally in the lower right-hand corner of the sheet.

The readability of construction drawings is dependent on the clarity and legibility of the graphic technique. Variations in line thickness (referred to as *line weight*) and line type (e.g., solid, dashed, or dotted) are used to indicate a hierarchy of importance, a change in level, or a change in material. The organization of the information also contributes to the readability of a drawing. Development of a consistent, clear lettering style will add to the quality of the drawing.

EXERCISES

5.1 Develop a sketch book catalog of different types of grade change devices.

5.2 Evaluate a variety of stairs and ramps in terms of walking comfort. Measure risers, treads, and slopes to determine comfortable (and uncomfortable) relationships.

The exercises in Chapter 3 provided all the criteria necessary to plot proposed contour lines. The following problems are less structured and, therefore, allow for the development of an approach to grading and the consideration of constraints that affect grading decisions.

5.3 A residence has been placed on a site as illustrated in Fig. 5.28. Using the finished floor elevations noted on the plan, regrade the site and indicate all proposed contour lines and spot elevations. The proposed exterior grades adjacent to the building cannot be less than 0.5 ft nor more than 1.0 ft below the established FFE, except at the garage door.

5.4 Several buildings and a parking lot have been placed on a site as shown in Fig. 5.29. Establish the FFE for each building, regrade the site, and indicate all proposed contour lines and spot elevations. The proposed exterior grades adjacent to the building cannot be less than 0.5 ft nor more than 1.5 ft below the established FFE at each building. The curb along the driveway and parking lot is 0.5 ft high.

5.5 As illustrated in Fig 5.30, six townhouse units have been placed on a sloping site. The architectural unit has been designed to accommodate the cross-slope by providing the following features:

a. The rear entry is 4.0 ft above the front entry elevation.

62   Grading Design and Process

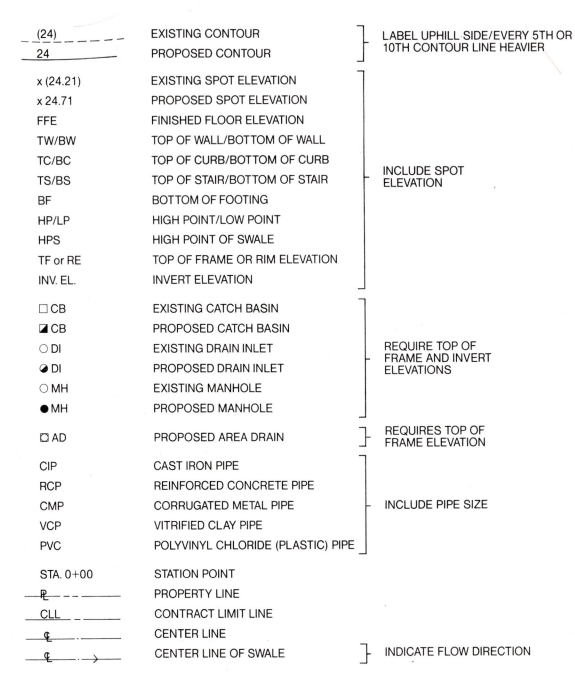

FIGURE 5.27. Typical Grading Plan Symbols and Abbreviations.

b. Each *pair* of units must have the same FFE; however, elevations of the units may change a maximum of 4.0 ft up or down at the fire walls.

c. The exterior grade at the front entry may equal the FFE; however, the exterior grade at the rear entry must be 0.5 ft below the rear FFE.

Regrade the site, indicating all proposed contour lines and spot elevations. Provide a usable outdoor living space for a distance of 8 to 10 ft at the rear of each unit.

5.6 This exercise deals with four attached dwelling units, walks, parking lot, and entry drive as shown in Fig. 5.31. The FFE must be the same for all units, and the rear entrances must be 9.0 ft below the front entrance elevations. There should be a 10.0-ft by 10.0-ft usable area adjacent to the rear entrance of each unit. The finished exterior grade at each entrance should be 2 in. below the proposed FFE. Grade-change devices (steps and walls) may be necessary to solve this problem. The existing contour lines are shown at 2-ft intervals. However, the proposed contour lines should be indicated at 1-ft intervals.

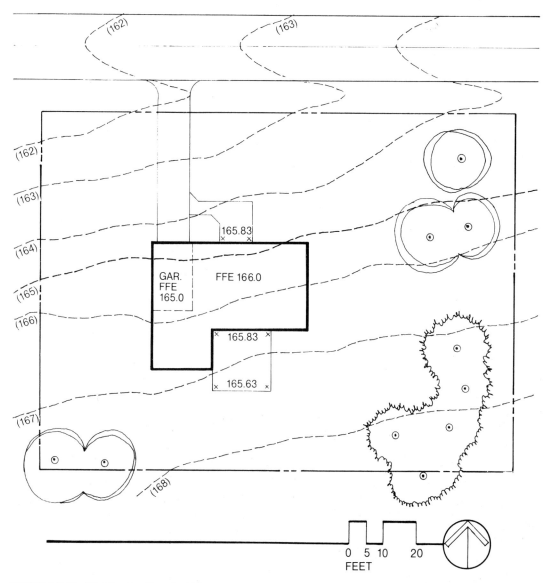

FIGURE 5.28. Site Plan for Exercise 5.3.

64 Grading Design and Process

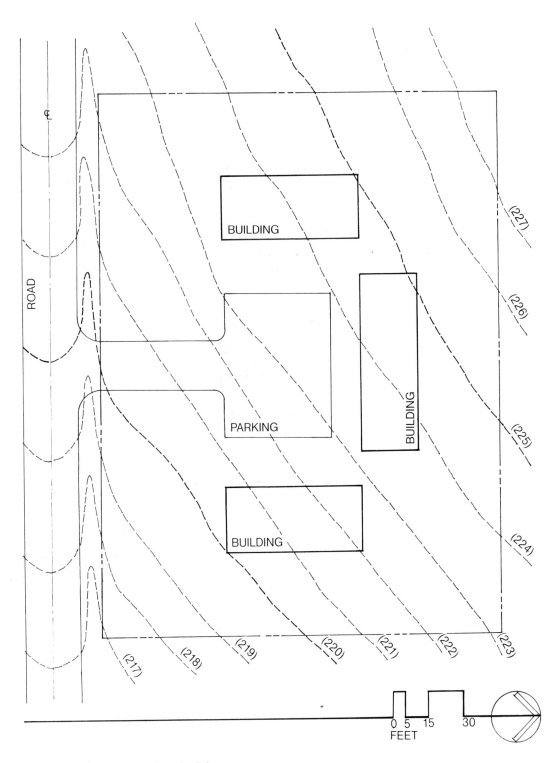

FIGURE 5.29. Site Plan for Exercise 5.4.

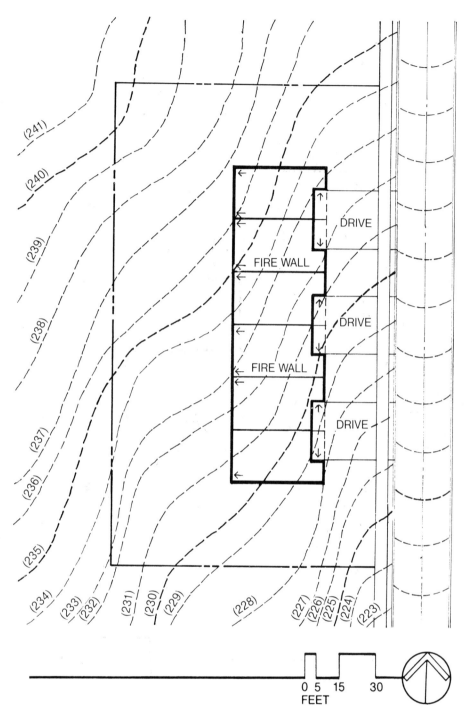

FIGURE 5.30. Site Plan for Exercise 5.5.

66 Grading Design and Process

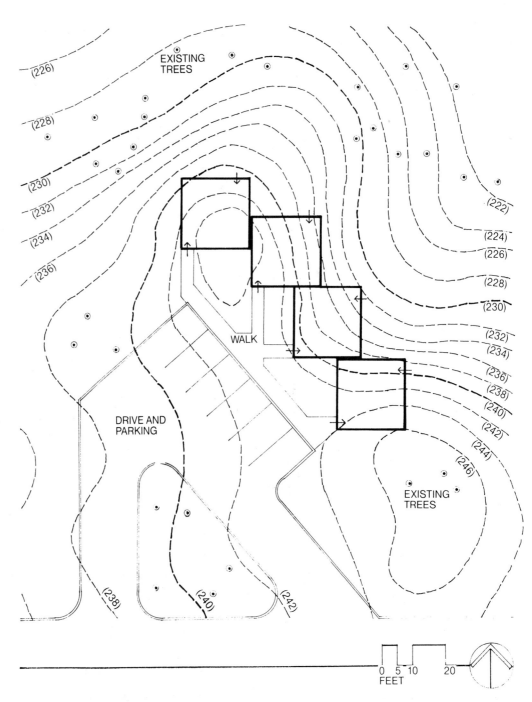

FIGURE 5.31. Site Plan for Exercise 5.6.

6
Grading and Landform Design: Case Studies

INTRODUCTION

As pointed out in the previous chapter, an understanding of the way landform and grading can be used aesthetically or spatially is an essential aspect of a site planner's design vocabulary. Landform design is more than just moving earth; it requires a clear set of goals and objectives to achieve a desired intent or result. As discussed in Chapter 5, landform design, including grade-change devices, such as walls, steps, and ramps, can be used to lead or direct us through a landscape, to create a sequence of experiences and spaces, such as an exposed condition on the top of a knoll or an enclosed feeling created by an amphitheater or bowl-like form. Landform design, used in conjunction with planting design, provides the basis for our visual and spatial perceptions of the landscape. In this chapter, several case studies are presented which demonstrate how landform and grading can be used to establish or reinforce the fundamental concept of a design.

EARTHWORKS PARK[1]

Earthworks Park in Kent, Washington, demonstrates how grading can be used to solve a pragmatic storm runoff and erosion problem while creating sculptural landforms and a park area for passive recreation. The design and reclamation of the Mill Creek Canyon site were undertaken to alleviate erosion and flooding problems as part of a public art project entitled "Earthworks: Land Reclamation as Sculpture" sponsored by the King County Arts Commission and King County Department of Public Works during the 1970s.

The park is located at the lower end of a 1500-acre drainage basin. Approximately 460 cubic feet per second (cfs) of storm runoff would flow through the site during a 100-year storm. The goal of the storm water management project was to reduce the discharge so that it would not exceed 100 cfs for a 100-year storm. To meet this goal, a large detention basin was created by constructing an earth dam across the steeply walled valley formed by Mill Creek. The resulting detention basin has a water storage capacity of 652,000 ft^3 (approximately 15 acre-feet).[2] (A discussion of storm water flow rates and storage is presented in Chapters 8 and 9.)

A series of abstract circular forms consisting of ring-shaped and cone-shaped mounds were created in the area of the park on the upstream side of the dam. The sculptural landforms establish the overall spatial and visual character of the park (Fig. 6.1). These landforms reflect an earlier earth sculpture work by Bayer, The Grass Mound, in Aspen, Colorado. The role of the park as a storm water management facility is visually reinforced by a number of design elements. A circular pond containing an inner grass ring (Fig. 6.2) retains water most of the year, thus providing a sense of the storm water detention function of the larger park landscape. A small stream passes under the bridge shown in Fig. 6.3 during dry periods; however, the capacity to accommodate much greater volumes of water is made readily apparent to the park user by the scale of the bridge structure. A ring-shaped mound provides a more enclosed space within the larger landscape. Retaining walls are used to define the entry into this smaller space (Fig. 6.4). The three primary sculptural elements (ring, bridge, and circular pond) are physically and visually linked by the stream running through the park.

---

[1]Designer: Herbert Bayer; Engineers: URS Engineers.

[2]Statistics from "Mill Creek Canyon," Kent Parks and Recreation Department, Kent, Washington.

68 Grading and Landform Design: Case Studies

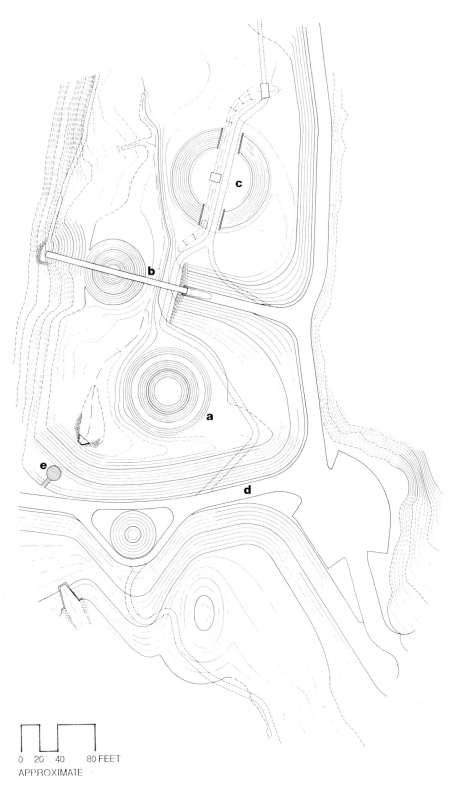

FIGURE 6.1. Site Plan of Earthworks Park. a. Circular pond. b. Bridge structure. c. Ring-shaped mound. d. Embankment. e. Spillway structure.

Earthworks Park 69

FIGURE 6.2. Circular Pond with Inner Grass Ring.

FIGURE 6.3. Bridge Structure.

FIGURE 6.4. Ring-Shaped Mound with Retaining Walls.

A set of access stairs at one of the valley walls is an interesting detail. Rather than switchback or scissor-type stairs, which would psychologically reduce the apparent height of the valley wall, the stairs ascend directly up the slope (Fig. 6.5). This technique accentuates the steepness of the landform. A small platform provides a resting and vantage point approximately halfway up the stairs. The placement of the platform next to the stairs is important to note, since it does not interrupt the steep character created by the line of the stair structure (Fig. 6.6).

GASWORKS PARK[3]

The design of Gasworks Park in Seattle, Washington, has been much publicized as a model for recycling abandoned industrial sites for recreational uses. The gasworks facilities on the site were retained as an example of industrial archeology and as a giant play sculpture for children and adults. The site, dramatically located on Lake Union, commands an excellent view of downtown Seattle. Landform was incorporated into the design of the park to take advantage of the view and waterfront location.

A knoll, called the Great Mound, was created at the water's edge to provide a vantage point for overlooking the park, Lake Union, and the city skyline. In a preliminary design scheme, a path spiraled up the knoll on the waterfront side of the landform. With this layout the view, which would be one of the primary reasons to climb to the top, would immediately be revealed to the park user and there would be little incentive to entice the user to climb to the top of the knoll. The path layout was refined to provide a choice. One route creates a series of switchbacks which progress up the landform on the side away from the

FIGURE 6.5. Access Stair at Valley Wall.

[3]Landscape Architect: Richard Haag.

FIGURE 6.6. Overlook Platform at Access Stair.

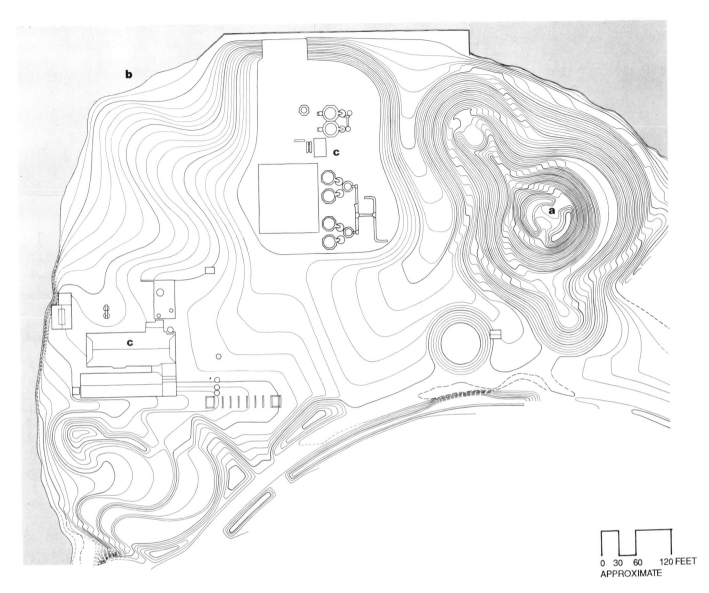

FIGURE 6.7. Site Plan of Gasworks Park. a. Great Mound. b. Lake Union. c. Industrial structures.

FIGURE 6.8. Switchback Paths Traverse the Landward Slope of the Great Mound.

water (Fig. 6.8). As a result of this configuration, the landform conceals the view as the pedestrian climbs up the hillside and adds to the sense of anticipation and arrival once at the top. The second route cuts across the waterfront side of the Great Mound and leads to the shore of Lake Union as well as to the top of the knoll (Fig. 6.9). Berms placed at the top of the Great Mound create a craterlike form, define the entrances to the summit where a sundial is located, and provide a sense of containment (Fig. 6.10).

FIGURE 6.9. Path along Waterfront Side of the Great Mound.

FIGURE 6.10. Summit of the Great Mound. Berms form a small craterlike form which creates a sense of containment at the summit.

OLYMPIC PARK[4]

Olympic Park is perhaps one of the most outstanding large-scale landform designs of the twentieth century. The design concept for the park is strongly defined through the use of grading and landform (Fig. 6.11). The site selected for the 1972 Summer Olympics had been used as the dumping area for the rubble cleared from Munich, Germany, after the bombings of World War II. A decision was made to incorporate this material into the design rather than remove it from the site.

The design scheme for the Olympic athletic facilities and surrounding park is an excellent example of how a complementary relationship can be achieved between architecture and landscape architecture. There are three major components to the design: the structures for the sports facilities, a lake, and a "mountain." Conceptually these elements are combined to create an illusion of a valleylike landscape. The walls of the valley are defined by landform (the Olympic "mountain") on one side and by architecture (the tensile structures of the stadium, sports hall, and swimming pool) on the other. A lake establishes the valley floor between these major elements (Fig. 6.12).

The large landform which creates the Olympic mountain is approximately 197 ft high at its highest point. It is traversed by an extensive system of walks and paths which, through changes in width, materials, steepness, and alignment, create a variety of experiences, from highly exposed to extremely intimate, for the pedestrian (Fig. 6.13). Because of its size and uniqueness, particularly within the context of an urban setting, the park also provides a variety of recreational opportunities from skiing and sledding to flying model gliders (Fig. 6.14). Although a grading plan was developed for the design of the mountain, it should be noted that creating a landform of this magnitude required a great deal of on-site direction and supervision during the earth-moving stage of construction to achieve the desired landscape and sculptural character.

The tentlike forms of the structures, particularly the stadium, create an illusion of mountains which reinforces the landscape concept for the park. One side of the sta-

---

[4]Architects: Günther Behnish & Partners; Engineer: Frei Otto; Landscape Architect: Günther Grzimek.

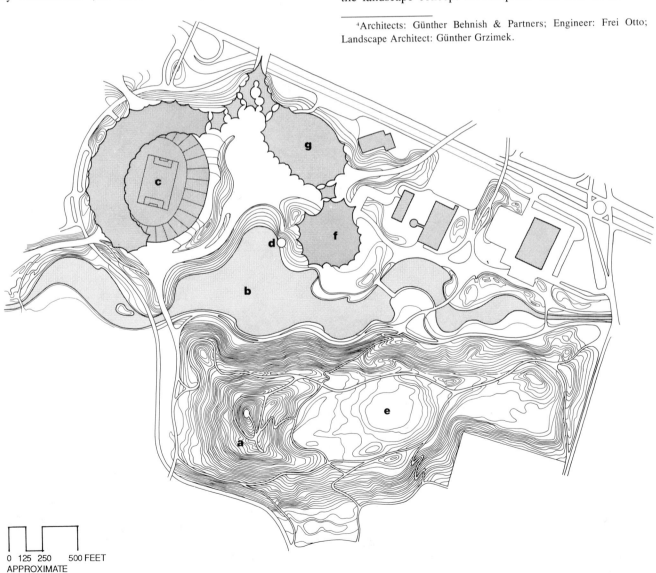

FIGURE 6.11. Plan of Olympic Park. a. Olympic mountain. b. Lake. c. Stadium. d. Amphitheater. e. Plateau/upper meadow. f. Pool. g. Sports arena.

74 Grading and Landform Design: Case Studies

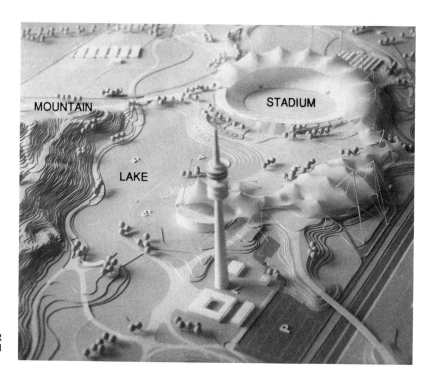

FIGURE 6.12. View of Site Model Illustrating the Valleylike Form Created by Landform and Architecture.

FIGURE 6.13. View of Olympic Mountain. Low height of slope planting enhances the perceived size of the mountain landform.

FIGURE 6.14. Wind Currents Created by Landform Make Olympic Park an Excellent Place for Flying Model Gliders.

dium has been tucked into the landform to reduce the apparent size of the structure so that it does not overpower the park landscape. From the park, the stadium appears much smaller than its capacity of 84,000, since it seems to sit in a "bowl," rather than pop out of a flat plane, as is typical of many stadiums (Figs. 6.15, 6.16).

As discussed earlier, landform design and planting design should be coordinated to produce a unified design concept. Planting on the Olympic mountain is an excellent example of this complementary relationship. Perceived scale of this artificially created mountainlike landscape was an important criterion in developing a planting scheme. The slopes which define one wall of the valley space are treated, for the most part, as a two-dimensional canvas on which plant material has been applied, since the steep slopes are perceived as an edge or boundary rather than a space which can be traversed (see Fig. 6.13). Small-scale plant material was selected, since it was felt that large-scale material (i.e., shade or canopy trees) would diminish the perceived height and overall visual impact of the landform. The grouping, scale, and texture of the plant material also create a somewhat false perspective, thus making the slopes appear larger than they are. The drama and sense of steepness may also have been lost if the slopes had been overplanted.

A plateau, consisting of a large, open grass area, has been created as part of the mountainlike landform. The approach to the planting changes in this area, recognizing

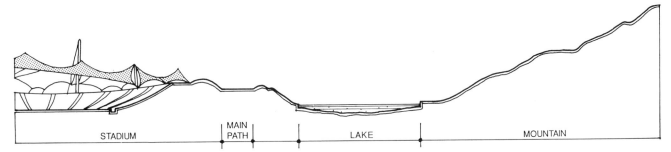

FIGURE 6.15. Section of Olympic Stadium, Lake, and Landform.

FIGURE 6.16. View across Lake Illustrates the Low Profile of Olympic Stadium.

that both vegetation and landform are needed to define this landscape space. Masses of large trees are used to help define the edges and spatial limits at one end of the plateau, while the summit of the mountain contains the space at the other end. The various types of landform (slopes, mountain, plateau) together with the different planting design treatments add to the diversity of the park experience. The plateau also allows for a wide variety of activities and uses which could not otherwise be accommodated on the steep slopes (Fig. 6.17).

The height of the slope illustrated in Fig. 6.18 demonstrates an interesting phenomenon. The people sitting on the slope are highly visible to surrounding park users. They can be seen easily while they command a panoramic view of the valley space. However, since they are also removed from the mainstream of activity, a personal level of privacy is achieved within this highly exposed public landscape.

Landform has also been used to separate functions and activities. Large paved areas and wide walkways are needed to accommodate large crowds of people around the stadium. This extensive circulation system is separated from the lakefront and smaller-scale paths through the manipulation of landform (Fig. 6.19). A berm and steep slopes completely block the view of the stadium from the walk in Fig. 6.20. The terraces in Fig. 6.21 create an amphitheater effect and provide an informal gathering and sitting space between the stadium and the lake. The stairs are sensitively placed into the terraced slopes with a staggered arrangement emphasizing the curvilinear form of the terraces (Fig. 6.22). Had the stairs simply cut across the terraces in a straight line, the curved geometry would have been interrupted and the small grade change across the terraces would have become more obvious. This condition is opposite to the design of the stairs at Earthworks Park, which accentuates the grade change. The transition from steep slope (Fig. 6.20) to grass terraces (Fig. 6.21) to a formally defined amphitheater (Fig. 5.9b) provides an interesting example of how a variety of grading techniques can be integrated to form a unified landscape setting.

One intent for creating the Olympic mountain was to establish a strong visual connection between park and city. Standing on top of the mountain provides a view to downtown Munich as well as to surrounding urban neighborhoods while the mountain is highly visible from the surrounding areas and the clock tower of the city hall in the center of downtown.

FIGURE 6.17. Plateau Provides a Relatively Flat Area for More Active Play.

Olympic Park 77

FIGURE 6.18. People Sitting on Slope Are Highly Visible, yet Removed from the Mainstream of Activity in the Park.

FIGURE 6.19. Aerial View of Path System between Stadium and Lakefront.

FIGURE 6.20. View of Lakefront Path Illustrated in Fig. 6.19. Steep Slope Separates Lakefront from Main Walkway and Stadium.

FIGURE 6.21. Grass Terraces Provide a Transition between Lakefront and Large Paved Gathering Space outside Stadium.

FIGURE 6.22. Aerial View Illustrates the Staggered Progression Created by the Placement of the Terrace Stairs.

WESTPARK[5]

In 1978 a design competition was held to design a new park in Munich, Germany, which would also serve as the site for the 1983 International Garden Exhibition. There were 26 entries, all of which were influenced to some degree by the use of landform at Olympic Park, which had been completed six years earlier. The park site was primarily a flat, abandoned industrial area with very little existing vegetation. Most of the submissions treated the site as a table top and proposed that the new landforms be constructed above existing grade. One submission, the one which was selected, recognized that Munich sits on a glacial lake bed of sand and gravel which is an excellent, stable construction material for excavating and filling. The selected entry created landform by excavating the soil and using the excavated material to construct the new landform. This solution maximized the degree to which landform could be created in a very economic manner. Approximately 2 million cubic yards (yd^3) of earth were moved in molding and creating this park landscape (Fig. 6.23).

The fundamental concept for the park was to recreate a sense of the regional landscape within the city. An abstracted fore-Alpine landscape was developed with a series of valleys as the basic theme. One main valley of varying widths moves through the center of the park and serves to unify the entire park landscape. Ridgelike landforms are used along the edges of the park to block noise and views from adjacent streets and highways and to heighten the natural experience within the park by blocking views to the city. Within the park conical landforms serve as abstracted mountains which provide vantage points to view the park, the city skyline, and, on clear days, the Alps located to the south of Munich. The oak trees pictured in Fig. 6.24, the only significant existing vegetation, provide a bench mark in relation to existing grade on the site. The magnitude of earth moving and the scale of the main valley which was constructed can be seen in these photographs.

An interesting aspect of Westpark is that it is divided into two parcels by a major city thoroughfare, as seen in the site plan in Fig. 6.23. The intent of the design was to make the park feel like one unified site rather than two separate areas. Budget restrictions, however, limited the size of the bridge which could connect the two parcels. A graceful connection was achieved by two large valley spaces which sweep up to meet the bridge on each side. The length of the bridge is relatively short in relation to the size of the valley spaces. While crossing the bridge the pedestrian's view is focused on the expanse of these spaces; thus the landscape becomes the emphasis of the experience and not the road which is being crossed. As a result, the pedestrian moves from one section of the park to the other in a flowing movement without a noticeable break, thereby the park is tied together successfully as a unified spatial experience (Fig. 6.25).

There are a number of grading details which add to the overall success of the design of Westpark. The treatment around the oak trees, where the landform meets the top of the seat wall, brings the foreground closer to the viewer and increases the amount of landscape perceived by viewers both into and out of the space (Fig. 6.26). Notice that

[5]Landscape Architect: Peter Kluska.

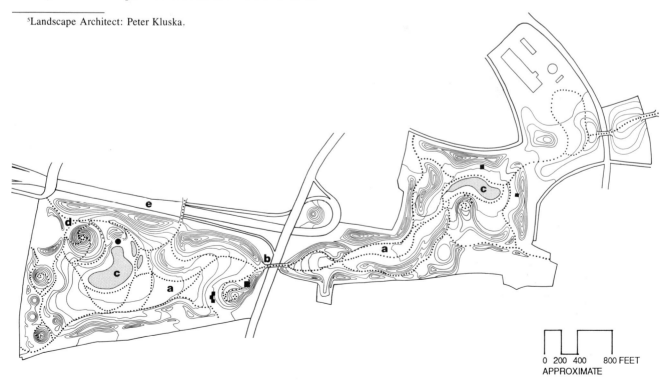

FIGURE 6.23. Site Plan of Westpark. a. Main valley. b. Bridge. c. Lakes. d. Conical landform (pictured in Fig. 6.28). e. Highway.

80   Grading and Landform Design: Case Studies

(a)

(b)

FIGURE 6.24. Large Existing Oak Trees Serve as a Bench Mark for the Original Ground Level of the Site. a. During construction. b. After completion of the park.

Westpark 81

FIGURE 6.25. Sequence of Views to and from Bridge Which Connects the Two Portions of the Park.

a. View to bridge from eastern portion.

b. View from bridge to large meadow space of western portion. Arch serves as a landmark, gateway, and frame.

c. View to bridge from western portion.

(a)

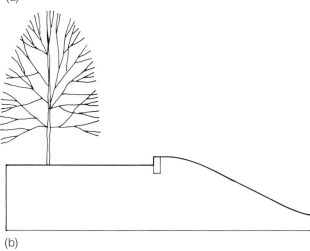

(b)

FIGURE 6.26. Grading at Oak Trees. a. View of wall and slope. b. Section.

(a)

the top of the slope is rounded, conforming to the principles presented in Examples 3.5 and 3.6 in Chapter 3. A similar condition can be seen in Fig. 6.27. Here, however, the earth has been slightly mounded to add to the sense of enclosure for the amphitheater space and to conceal the amphitheater from the surrounding landscape.

Spiral walks are used to traverse the steep conelike landforms. The detailing has been handled in several ways, including ramps, stairs, and a combination of ramps and stairs. These techniques are illustrated in the previous chapter in Figs. 5.10b and 5.13c. The tops of the conical forms have been treated in a variety of ways, one of which is illustrated in Fig. 6.29. This treatment demonstrates how subtle grade changes can be used to define edges, enclose spaces, and create a variety of circulation experiences for the pedestrian.

An interesting comparison can be made between the grading design concepts employed at Westpark and Olympic Park, two major urban park spaces which are located only a few miles apart. Conceptually, Olympic Park places the user/observer either on a mountain in a highly exposed or visible situation or in a valley in which the landform or architecture becomes the focal point of the park experience. Westpark places the user in a valley where he or she is immersed in an abstracted, natural landscape. Glimpses of the city are obtained from the ridge lines along the edges of the park or from vantage points provided by the conical landforms. However, these vantage points are much less exposed and visible than the mountain at Olympic Park. Both parks use the valley to establish the organizing spatial framework for their respective landscapes. However, in Westpark, the primary experience is a sequence of landscape spaces, whereas in Olympic Park the primary experience is the visual domination of the mountain and athletic facility structures. The latter is a more sculptural and object-oriented experience, whereas the former relies more on the kinesthetic experience of moving through a sequence of landscape spaces.

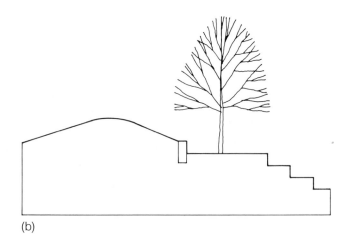

(b)

FIGURE 6.27. Grading at Amphitheater. a. View of berm and wall. b. Section.

FIGURE 6.28. View of conelike landform which creates an illusion of a mountain landscape. Notice how the paths spiral up the landform to reduce the steepness of the slope.

FIGURE 6.29. Sitting Area at Summit of Landform Shown in Fig. 6.28.

EXERCISES

6.1 There are many excellent examples of grading and design. Select sites within your region and conduct case study analyses which examine design intent and details and techniques used to reinforce the design objectives. Document your analyses through the use of plans, sections, axonometrics, diagrams, and photographs.

7
Storm Water Management

STORM RUNOFF

Storm or surface runoff is storm water that moves on the ground by gravity and flows into streams, rivers, ponds, lakes, and oceans. For impervious surfaces like pavements and roofs, runoff occurs almost immediately. For pervious surfaces, the intensity of precipitation must exceed the infiltration rate: that is, the surface must be saturated before runoff will occur. *Subsurface runoff* is storm water that infiltrates and moves through the soil both horizontally and vertically. The rate of movement is influenced by soil permeability and usually is much slower than that of surface runoff.

As noted in Chapter 4, the acts of grading and controlling and managing storm water runoff are inextricably linked. Almost all site development projects result in the remolding and sculpting of the earth's surface as well as changes in surface character. These changes may significantly alter storm runoff patterns in terms of rates, volumes, and direction. Landscape architects and site planners must understand the consequences, if these changes are to be effected in a safe, appropriate, and ecologically sensitive manner. This chapter provides an introduction to basic management principles and techniques and potential problems caused by storm water runoff. The proper design of any management system requires an interdisciplinary approach, including professional expertise in ecology, engineering, hydrology, and landscape architecture.

It cannot be stressed strongly enough that all storm water management practices must be *site-, region-,* and *climate-specific.* In addition, legislative controls in terms of runoff, erosion, sedimentation, and water quality have become increasingly common from the local to the state and federal levels. It is imperative that any proposed site plan comply with appropriate regulatory requirements.

HYDROLOGIC CYCLE

The hydrologic cycle is a natural, dynamic process, which is diagrammatically illustrated in Fig. 7.1. Within a natural system, changes such as creation of river valleys, redirection of a stream channel, or erosion and sedimentation usually occur slowly over long periods. Generally site development disrupts the natural hydrologic cycle by accelerating runoff and reducing the proportion of precipitation which infiltrates the ground or is taken up by vegetation. It is the role of the site planner to minimize, mitigate, or ameliorate these disruptions to the natural system through appropriate storm water management and sound land use practices.

NATURE OF THE PROBLEM

Urbanization has a profound impact on existing natural and constructed drainage systems. Development typically results in an increased amount of impervious surfaces, such as roofs, streets, parking lots, and sidewalks. The consequences of these surface changes are numerous but are primarily rooted in the fact that developed sites lose much of their natural storm water storage capacity. The loss of vegetation, organic litter, and changes in surface characteristics such as roughness and perviousness result in the rapid conversion of rainfall to storm water runoff. Often the increased rate and volume of runoff become too great for existing drainage systems to handle. In order to accommodate the increases, drainage systems are structurally altered through the use of curbs, gutters, channels, and storm sewer pipes to direct and convey runoff away from developed areas.

Several environmental impacts may result from changes in the storm drainage pattern, including increased flood potential due to increases in peak flow rates, decreased

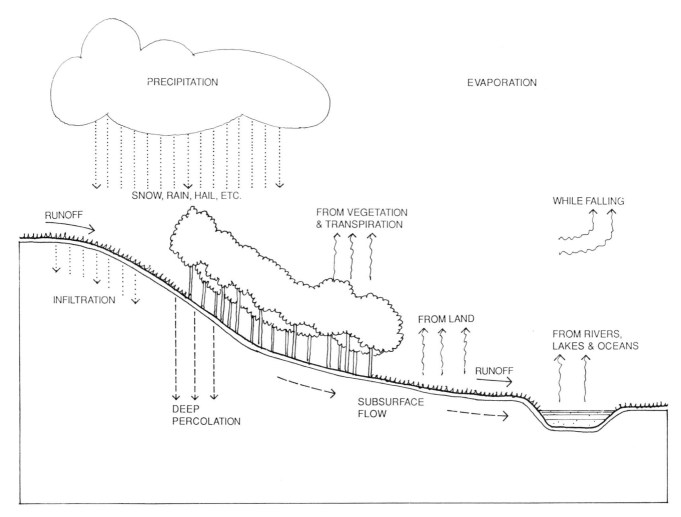

FIGURE 7.1. Hydrologic Cycle.

groundwater supply caused by reduced infiltration, increased soil erosion and sedimentation brought about by greater runoff volumes and velocities, increased petrochemical pollution from street and highway runoff, and addition of salt and sand to winter runoff in colder regions. Addressing these and other issues in the design and implementation phases will result in a more environmentally responsive management system.

Hydrologic Changes

There are a variety of changes in stream hydrology which result from development. Peak discharges, which can be as much as two to five times the predevelopment rate, generally increase the frequency and severity of flooding. Development which had once been considered above flood level may become subject to flood damage. In addition to increased rates of runoff, volume of runoff may also be increased, as a result of reduced infiltration and storm water storage capacity. Higher runoff velocities, which also reduce the time for peak discharge to reach a stream or drainage channel, result from smoother surfaces such as pavements which create less friction to slow runoff flow. Increased velocities and/or shorter overland travel times

also provide less opportunity for infiltration. Higher velocities coupled with increased imperviousness may also reduce stream flow during extended dry periods caused by reduced infiltration. Groundwater which would normally be recharged during wet periods and released slowly from soil during dry periods is lost as surface runoff (Fig. 7.2).

Stream geometry also changes. Streams are widened by increased volume and velocity, which results in increased stream bank erosion. Usually the stream bank is undercut, destabilizing vegetation and, in turn, exacerbating the erosion problem. Eroded material is deposited in stream channels as sediment which reduces stream flow capacity. Flood elevations are raised; as noted, this phenomenon increases the extent of the area at risk of flood damage.

The quality of storm water is also degraded as a result of development. Pollutants are accumulated on paved surfaces and are flushed from these surfaces during a rainstorm. Not only do developed or urbanized landscapes increase the ease with which pollutants can be collected and concentrated, but they also increase their sources. Contaminants may be released through corrosion, decay, oil and fuel leaks, and leaching or wearing away of construction materials and coatings, brake linings, tires, and catalytic converters. Developed areas, as well as

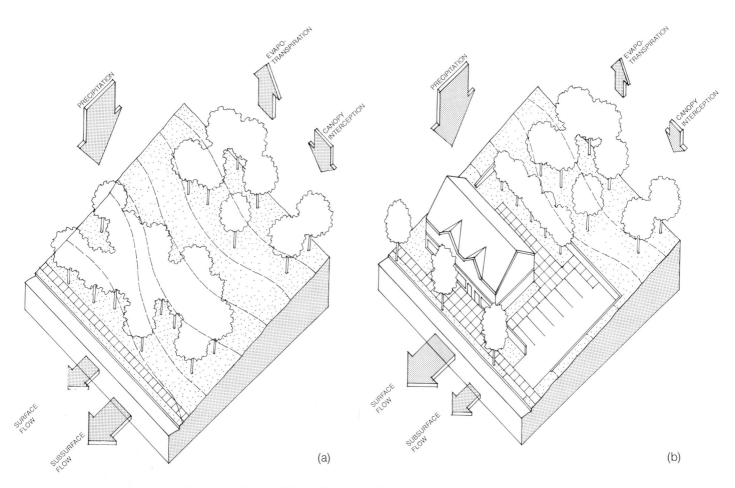

FIGURE 7.2. Relative Water Balance. a. Undeveloped site. b. Developed site.

agricultural areas, contribute herbicides, pesticides, and fertilizers, which stimulate algal growth and reduce the availability of oxygen in water. Fallen leaves and animal droppings which would normally decompose in undeveloped or low-density settings are more easily washed from paved surfaces, further contributing to nutrient and bacterial levels in streams, ponds, and lakes.

Changes in stream hydrology and geometry, combined with reduced water quality, decrease the value of aquatic, stream bank, and floodplain habitats. Not only are stream channels and flows altered and pollution levels raised, but conditions are further exacerbated by low summer flows and higher water temperatures.

MANAGEMENT PHILOSOPHY

Historically, the primary concern in dealing with storm water runoff was to remove it as quickly as possible from a developed site to maximize local convenience and protection. Traditionally this was accomplished by conveying runoff by storm sewers, swales, gutters, and channels to the nearest water body, usually a stream or river. Little consideration was given to potential off-site impacts. This practice has had a cumulative effect on environmental and water quality.

More recent storm water management practices have recognized the need for controlling off-site impacts caused by increased runoff volumes and peak discharge rates. The objectives of these revised practices have been twofold: first, to reduce downstream flooding through the use of detention facilities which store and release runoff at a controlled rate and second, to reduce flooding damage by restricting floodplain development. Most legislation at either the state or local level requires that the peak rate of runoff after development not exceed the rate before development. In some cases the peak development rate must even be less than the predevelopment rate. However, such practices, when executed on a site-by-site basis, may in fact still not solve regional flooding problems and fail to address water quality and habitat issues.

The most recent philosophy with regard to storm water management is to develop a comprehensive, integrated approach which addresses water quality in addition to volume and rate of runoff. One of the primary management objectives is to deal with runoff on-site rather than transporting the problems off-site. A basic objective for any site design should be to minimize hydrologic problems by preserving and maintaining the predevelopment drainage patterns to the greatest extent possible. The preservation of existing drainage patterns and the use of appropriate management techniques, including detention, storage, and infiltration, will normally reduce the cost of the proposed drainage systems by minimizing the need for piping and drainage structures.

Measures which have been developed to control, store, and/or treat storm water runoff from developed areas for the purpose of reducing flooding or removing pollutants while maintaining or enhancing environmental quality are referred to as best management practices (BMPs). The goal of such practices is to control nonpoint source pollution while providing effective storm water management. A BMP for a specific site should be designed to control runoff, maximize pollutant removal, and integrate with the natural and built landscape with consideration for maintenance requirements, costs, and responsibilities. Through proper planning and good design, storm water management facilities can serve multiple uses, provide community and aesthetic amenities, create safe environments, and reduce development costs.

PRINCIPLES AND TECHNIQUES

BMP selection criteria include storm water management objectives, water quality objectives, and appropriateness to specific site conditions, including climate, soils, topography, proposed and existing land use, and surface cover. Integration of specific management devices into the existing landscape and proposed development poses a challenge for site planners and landscape architects. The design of a best management practice has several components: The nontechnical aspects include aesthetics, site suitability and appropriateness, safety, cost, maintenance, and multiple use. These aspects will most likely establish the physical framework for the overall drainage concept. The technical aspects include understanding of the existing hydrologic characteristics, and engineering for the proposed system which will establish size, storage capacity, discharge, and infiltration rates. Control measures may be used singly or in combination, depending on the size and character of the existing site and the nature of the proposed development.

Best management practices commonly used include, but are not limited to, wet ponds, detention facilities, infiltration facilities, and water quality basins, or a combination of these practices. Typical design characteristics and capabilities for each of these are briefly described.

Wet Ponds

Retention (or *wet*) *ponds* are basins which contain a permanent pool of water. This control measure, through careful planning and design, can serve multiple purposes, including storm water management, pollutant removal, habitat improvement, and aesthetic enhancement (Fig. 7.3). Potential benefits derived from wet ponds may include increased property values, recreational opportunities, and creation of wildlife habitat. The configuration and edge treatment of retention ponds may be designed to appear refined, naturalized, or wild. Possible disadvantages include safety problems, algal bloom, offensive odors, mosquitoes, and need for maintenance and sediment removal.

The following features, summarized in Fig. 7.4, should be considered when designing a wet pond:

1. Maximize flow length between inlet and outlet. This extends the flow path before the water exits, thus increasing the time for sediment and pollutant settlement. A suggested width to length ratio is 3:1. If this ratio cannot be achieved, baffles can be used to break up flow and lengthen the flow path between inlet and outlet.
2. The pond should expand gradually in the direction of flow. Water entering the pond spreads out and uniformly displaces the existing water, thus preventing the creation of "dead zones." Generally an irregular shoreline is preferred when a more natural appearance is desired.

(a)

(b)

FIGURE 7.3. Examples of Retention Basins Used as Site Amenities. a. Pond is an attractive feature at the entrance to a residential development. Jets help to aerate the water to reduce algal growth. b. The stone edging provides a more refined appearance for this pond used to enhance the setting of a corporate campus.

FIGURE 7.4. Retention Basin (Wet Pond).

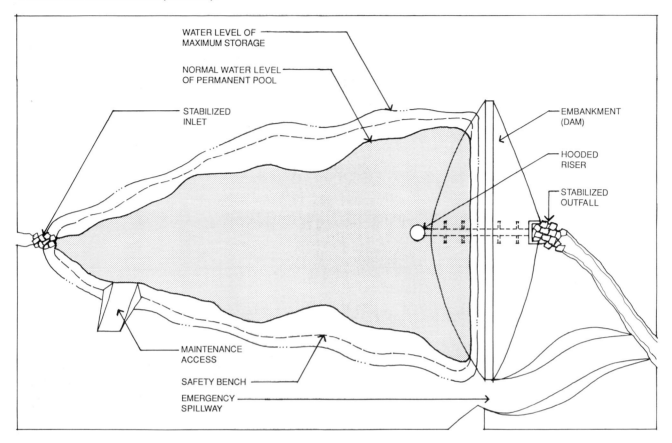

a. Plan.

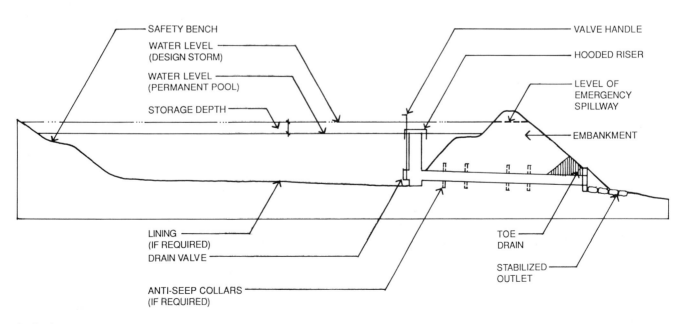

b. Section.

FIGURE 7.4. *continued.*

c. Pipe inlet with headwall and apron stabilized with riprap to minimize erosion.

d. Swale directs runoff to this retention basin. Gabions are used to stabilize the inlet point of the swale.

3. Pool depth should be between four and eight feet. Depths less than four feet can result in elevated water temperatures and resuspension of sediment produced by surface disturbance. A level safety bench at least 10 feet wide by 1 foot deep should be provided around the perimeter of the pond to reduce potential safety problems. Where groundwater quality is a concern, the pond should be lined and the basin depth should be above the seasonal high water table to prevent mixing of groundwater and runoff. Obviously under these circumstances, wet ponds cannot be used to recharge groundwater.

Detention Facilities

Detention facilities, or *dry basins,* are used as a means of controlling peak discharge rates through the temporary storage of storm runoff. Outflow rates are set at or below predevelopment rates, and flow is metered out of the basin until no water remains. Detention basins may work effectively in reducing downstream flooding and stream bank erosion, depending on the quantity of storm water detained and release rates, but may not be very effective in enhancing storm water quality. Detention ponds which extend temporary storage time allow for more effective removal of particulate pollutants and provide a technique for improving water quality.

On-site detention can be accomplished in a variety of ways. Surface techniques use ponds, basins, and paved areas; subsurface techniques employ dry wells, porous fill, oversized drainage structures, and cisterns or tanks. Subsurface techniques may be costly but reduce land consumption and may be appropriate on very tight sites where space for use of surface techniques is not available.

The design characteristics of detention basins are similar to those of retention facilities in many ways; however, additional features illustrated in Fig. 7.5 are required to increase potential use, enhance appearance, and improve maintainability. Detention basins can be designed to fit into a variety of locations, including lawns, playfields,

FIGURE 7.5. Detention Basin (Dry Pond).

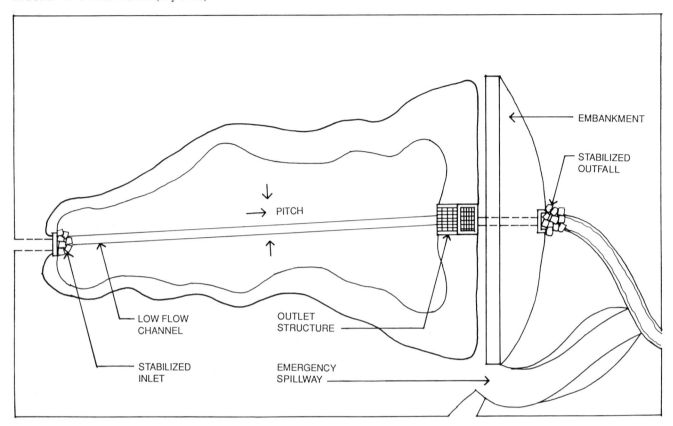

a. Plan.

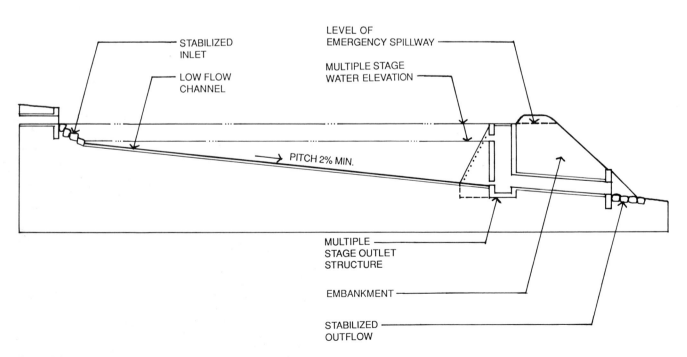

b. Section.

FIGURE 7.5. *continued.*

c. Large detention basin for a residential development.

d. Detailed view of inlet pipe and headwall pictured in Fig. 7.5c. Note the low-flow channel and the use of riprap for soil stabilization.

e. Multiple-stage outlet structure and concrete-lined emergency spillway for detention basin also shown in Fig. 7.5c.

open spaces, swales, parking lots, and court sport areas. Where multiple use is desired, provisions must be made to prevent standing water and minimize saturated soil conditions. These include the use of a low flow channel, positive drainage toward the channel and outlet, subsurface drainage for play fields, and a separate sediment basin upstream from the primary basin, easily accessible for maintenance purposes.

1. The length to width ratio is less critical for a detention basin than for a wet pond, but an elongated form is still preferred. Flow length between inlet and outlet should be maximized, and short-circuiting of flow should be prevented.
2. The side slopes of the basin should not be steeper than 3:1. The floor of the basin should have a 2% minimum slope toward the outlet to ensure positive drainage. An access way at least 10 feet wide with a slope of 5:1 or less should be provided for maintenance equipment.
3. A low-flow channel should be provided to reduce drying time and improve usability.

Parking lot detention areas may be an inexpensive way to control runoff, reduce storm sewer pipe sizes, and reduce erosion. Collecting runoff in a parking lot may cause inconvenience to traffic and pedestrians, but if the collection area is properly located, such conflicts can be minimized. Again, on small sites with little area available for detention or retention basins, parking lot detention may be an appropriate alternative.

Safety

Measures should be taken to reduce safety hazards which may be created by retention and detention ponds located in populated areas. Safety issues are related to access, large volumes of flowing water, constrictions created by pipes and culverts, and the intermittent nature of storm water storage. Safety measures may include installing fencing, avoiding steep side slopes or sudden drops, minimizing constriction points, and covering outlets with properly designed grates. As a minimum, the construction of detention and retention basins must meet all applicable federal, state, and local regulations, including state dam safety regulations where appropriate.

Infiltration Facilities

Infiltration techniques are highly beneficial in that they can significantly reduce or eliminate surface runoff while replenishing groundwater, which supplies wetlands, streams, and wells. Other benefits of infiltration include reductions in downstream peak flows and soil settlement caused by depletion of groundwater, preservation of existing vegetation, and lowering of development costs by reducing the size of the storm sewer system needed. Preserving infiltration after development can be one of the most effective mechanisms in preventing adverse impacts to the surface water system. In terms of water quality, infiltration devices filter runoff through the soil layer, where a number of physical, chemical, and biological removal processes occur. However, infiltration practices should not be used to remove sediment or other particulates, since sedimentation will eventually clog the infiltration device and render it useless. Sediment should be removed by the use of vegetated filter strips or sediment traps before it enters an infiltration device.

There are practical limitations to the use of infiltration facilities. These include soil permeability rates, potential reduction in permeability rates over time, and potential groundwater contamination. Conditions which must be examined to determine the appropriateness of infiltration facilities include depth to groundwater, seasonal variation in groundwater levels, slope and direction of groundwater flow, soil permeability, vegetative cover, and quality of storm water runoff.

Infiltration facilities may be classified as surface systems such as basins, subsurface systems such as trenches, or porous pavement systems.

Infiltration Basins

An infiltration basin is a surface impoundment created by damming or excavating. The purpose of the basin is to store runoff for a storm of a specific frequency and duration (*design storm*) or specific volume temporarily so that the water will enter the soil over a specific period. These basins, in terms of performance and appearance, are very similar to detention basins, and, in fact, basins are often designed to combine infiltration and detention functions (Fig 7.6). In addition to the limitations mentioned, infiltration basins may require large areas, are not adaptable to multiple use, and have high rates of failure due to improper maintenance and installation.

Infiltration Trenches

An infiltration trench is an excavation backfilled with coarse aggregate stone. The voids between the aggregate material provide the volume for temporary storage of storm runoff. Runoff stored in the trench gradually infiltrates the surrounding soil. The surface of the trench may be covered with grass with a surface inlet, or with porous material such as sand, gravel, or stone. Infiltration trenches are appropriate for relatively small drainage areas. They are flexible systems which can be fit into underutilized or marginal areas of a site easily and, where soil and groundwater conditions allow, can be readily adapted to retrofitting an existing developed site. An observation well, such as a perforated polyvinyl chloride (PVC) pipe placed vertically in the trench, should be installed to monitor any change in infiltration rate periodically (Fig. 7.7).

Porous Pavement

Usually pavement creates an impervious surface which generates high rates and quantities of runoff. The objective of porous pavements is to increase perviousness while providing a stable, protective surface. There are a variety of

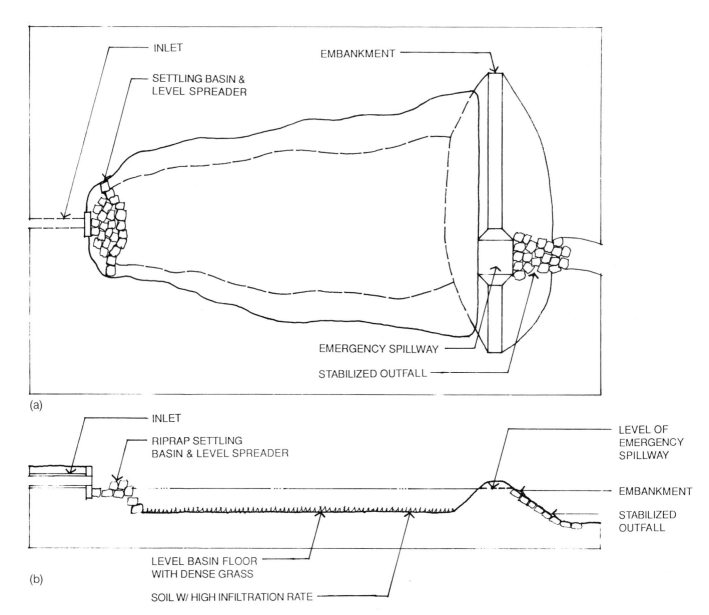

FIGURE 7.6. Infiltration Basin. a. Plan. b. Section.

porous pavement materials, including porous asphalt, open-gridded modular pavement, and pervious concrete.

Both porous asphalt and pervious concrete use an open graded aggregate mix which increases the interconnected pore or void space and allows water to percolate through the pavement (Fig. 7.8). Open-gridded modular pavement is typically constructed of a structural concrete framework with regularly spaced void areas which are filled with pervious material such as sod, sand, or gravel. Porous pavements are appropriate for such uses as low-volume roads, driveways, parking lots, bikeways, and emergency lanes.

Porous pavements should be used only where the subgrade soil conditions provide the proper permeability, depth to groundwater does not pose a problem, or contamination will not be caused by degraded storm water quality. It should be noted here that, in addition to use of porous pavements in infiltration situations, they can also be combined with subsurface storage facilities on densely developed sites.

The advantages and disadvantages of porous pavements are numerous. Advantages include generally higher recharge rates than natural conditions, as a result of less vegetative uptake; control of both rate and volume of runoff; reduced potential for swale and channel erosion; preservation of existing vegetation by maintaining proper soil moisture levels; reduced surface puddling; enhanced pollutant removal; and reduced construction costs through decreased infrastructure such as curbs, drains, and piping.

A major concern is limited experience with these systems among engineers and contractors. A high level of workmanship is required in preparing the subgrade, placing the base course, and installing the paving material if the system is to perform at the desired level. Another serious drawback is the potential for clogging if the pavement is improperly maintained or installed. Other disadvantages include potential for groundwater contamination, weakening of the subgrade in saturated conditions, and development of anaerobic conditions where storms are

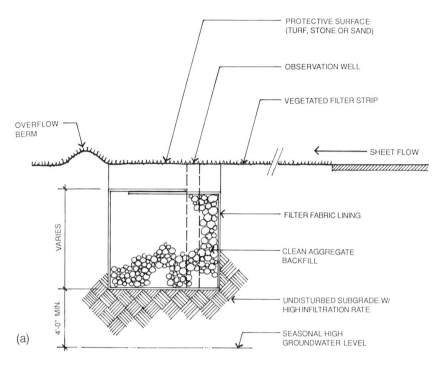

FIGURE 7.7. Infiltration Trench. a. Section. b. Sheet flow from parking lot is directed to stone-lined swale. Stone reduces runoff velocity, which enhances the potential for increased infiltration.

frequent. Also, in cold climates, sand and deicing salts cannot be used on porous pavements.

Water Quality Basins

Water quality amelioration is often included in the design of retention and detention basins. In retention (wet) basins additional volume is provided for the settling out of sediments. Detention (dry) basins may have a small orifice at the bottom of the outlet structure, so that runoff from frequent storms, which tends to flush contaminants into the pond, is retained for a prolonged period. The extended detention period resulting from the small orifice allows the contaminants to settle out and the pond to drain gradually.

Landscaping Practices

In and of themselves, landscape practices do not provide complete storm water management or water quality enhancement. However, several planting measures are an integral part of many BMPs; these employ vegetated swales, filter strips, basin landscaping, and urban forestry.

Vegetated swales provide an alternative to curbs and gutters and are commonly used for controlling storm water runoff, particularly in residential areas. Benefits derived from the use of swales include increased surface friction, resulting in decreased flow velocities and, consequently, increased times of concentration. The runoff-reducing performance of swales can be improved in combination with

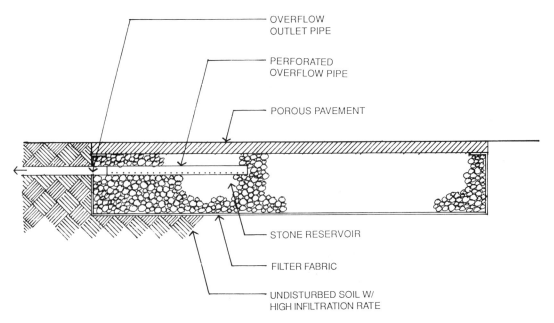

FIGURE 7.8. Porous Pavement.

other measures such as use of check dams and infiltration trenches. Swales have limited capacity and are subject to erosion if improperly designed or maintained.

Filter strips are devices which are typically placed adjacent to impervious surfaces to intercept overland sheet flow. Filter strips can lower velocities, increase time of concentration, improve infiltration, and contribute to groundwater recharge. They can be designed to fit into the overall site plan as screening, buffers, and spatial delineators. These strips must be properly designed to prevent concentration of runoff, which would eliminate their effectiveness. Grass filter strips should be used to trap sediment to protect the performance of infiltration trenches.

Landscaping is an essential component of all storm water basins. The design of retention, detention, and infiltration basins must include appropriate planting measures to ensure optimal performance. Benefits derived from properly designed planting schemes include slope stabilization, pollutant removal, aesthetic enhancement, and creation of wildlife habitat. Plant selection should respond to specific microclimatic and site conditions. In particular, it is important to recognize that there may be several different vegetative zones based on depth of water and different soil moisture conditions.

Urban forestry involves the preservation of existing trees and the planting of new trees during the development process. Benefits include the significant interception of precipitation by tree canopies and the increased water storage and infiltration capacity of the ground layer. The amount of runoff generated by areas landscaped with trees, shrubs, and ground covers may be as much as 30% to 50% less than that generated by lawn areas.

CONTROL OF EROSION AND SEDIMENTATION

The erosion and sedimentation process involves the detachment, transportation, and deposition of soil particles by the action of water, ice, wind, and gravity. Rainfall impact, flowing water, freezing and thawing, and wind dislodge soil particles; moving water and wind transport the particles and deposit them in a new location. The discussion which follows is limited to erosion caused by the force of falling and flowing water.

Erosion and sedimentation are critical issues in terms of storm water management, especially during construction, when exposed sites are particularly vulnerable. The rate of erosion for construction sites has been estimated to be 5 times that of agricultural land, 10 times that of pasture land, and 250 times that of forest land (Fig. 7.9). By volume, sediment is the largest nonpoint water pollutant. Examples of problems created by excessive erosion include filling of lakes, ponds, and wetlands; decreased channel capacity in streams and rivers; degraded animal and plant habitats; and increased potable water treatment costs. Many states and/or local jurisdictions require the submission of an erosion and sediment control plan before site plan approval is granted or a building permit is issued. It is the responsibility of the site planner to comply with all regulatory requirements.

Soil Erosion Factors

As shown in Fig. 7.10, four primary factors determine the potential for erosion: soil type, vegetative cover, topography, and climate. Each of these factors will be briefly discussed.

(a)

(b)

FIGURE 7.9. Examples of Severe Erosion on a Construction Site.
a. Concentrated runoff across bare soil results in gully erosion.
b. Sediment deposited on paved surfaces, where it can be easily washed into storm sewers or picked up and transported by car and truck tires.

Soil

The erodibility of a soil is determined by particle size and gradation, soil structure, permeability, and organic content. Soils with high silt and fine sand content have a high erodibility potential; soils with increased clay and organic content have a lower potential. Although clay does not erode as readily, once in suspension it does not settle out easily. Well-drained, well-sorted gravels and gravel-sand mixtures are highly permeable, are easily infiltrated, and therefore have the lowest erosion potential. Organic content enhances permeability and water absorption characteristics of a soil, thus decreasing erodibility. Various models exist for estimating erosion rates and evaluating various erosion and sedimentation management practices. The Universal Soil Loss Equation is perhaps the most widely known of these; however, a discussion of this topic is beyond the scope of this text.

Vegetative Cover

Vegetation prevents soil erosion in a variety of ways: First, it shields soil from raindrop impact. This serves to dissipate energy which would otherwise dislodge soil particles. Second, it retards runoff velocity through increased surface friction. Finally, plant root systems hold soil in place while increasing its water absorption capacity. The goal of most site development projects should be to retain as much existing vegetation as possible, particularly in vulnerable areas, such as steep slopes, poor soils, stream banks, and drainageways.

Topography

The length and steepness of slopes influence the amount and rate of storm water runoff. As the extent and gradient of slope increase, the amount, rate, and velocity of runoff increase, thereby increasing the potential for erosion. To reduce erosion caused by topographic conditions, avoid developing steep slopes, limit the length and gradient of proposed slopes, and protect disturbed slopes as quickly as possible. Slope orientation may influence the ability to reestablish a protective vegetative cover.

Climate

The frequency, intensity, and duration of rainstorms directly influence the amount of runoff generated on a particular site. Where possible, site construction should be scheduled during months in which low precipitation and low runoff are anticipated, although in some regions of the country this may be very difficult to predict. Scheduling construction to coincide with optimal seeding periods for a particular region would also promote the establishment of a cover crop for soil stabilization.

Erosion Control Principles

Appropriately planned erosion and sediment control measures which are properly installed and maintained from the initial phases through to the completion of a construction project can significantly reduce soil loss. The goal of

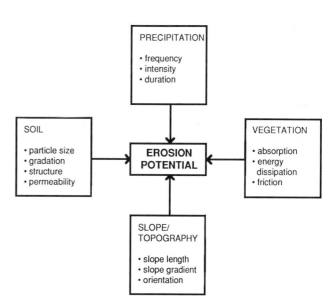

FIGURE 7.10. Factors Affecting Erosion Potential.

any plan should be to reduce soil detachment and transport. The following are basic principles which should be employed on any construction site to minimize soil erosion and sediment problems. Through proper planning, many of these control measures can become a permanent part of the storm water management system.

1. The area of disturbance and duration of exposure should be minimized. This can be accomplished by minimizing the developed area of a site, maintaining as much natural surface cover as possible, phasing construction, or stabilizing exposed areas through the use of mulch or a cover crop such as annual ryegrass.
2. Storm water runoff from upslope areas should be prevented from crossing disturbed or exposed soils. Diversion measures can be used effectively to redirect runoff. Where diversion is not possible, vegetative buffer strips should be established to reduce runoff velocities.
3. Storm water runoff generated on-site must be carefully controlled. Low gradients, short slopes, and preservation of existing vegetation, coupled with stabilization techniques such as sodding, seeding, or mulching, will help to reduce runoff velocities. In addition, runoff must be directed and routed to sediment control devices (traps, interceptors, silt fences, etc.) if these measures are to operate effectively.
4. Where erosion is unavoidable, the resulting sediment should be collected on-site. Sediment traps and basins, silt fences, check dams, and similar devices should be employed to prevent the off-site transport of eroded soil. Stabilized construction site access points should be used to reduce the tracking of soil from construction equipment tires onto adjacent streets or other paved surfaces.
5. A routine maintenance schedule is essential during construction if control measures are to function optimally. Situations such as silt fence or hay bale blowouts and filling of sediment traps with sediment must be tended to immediately.

Erosion and Sedimentation Control Measures

Control measures can be classified according to intent as summarized in Fig. 7.11. In terms of purpose, there are three ways in which control can be accomplished: soil stabilization, runoff control, and control and capture of sediment before it leaves the site. Methods, or particular control measures or devices, may be classified as vegetative or nonvegetative (structural). These preventive and collection techniques can be used alone or in combination.

Stabilization

Vegetative measures which provide soil stabilization include maintaining and protecting as much of the existing vegetation as possible; retaining existing topsoil for reuse; establishing new vegetative cover, for example, by seeding disturbed areas as quickly as possible; and selecting plants

	SOIL STABILIZATION	RUNOFF CONTROL	SEDIMENT CONTROL
CONTROL MEASURE	VEGETATIVE SOIL COVER	DIVERSIONS	SEDIMENT BASINS
		WATERWAYS	
	NON-VEGETATIVE SOIL COVER	OUTLET STABILIZATION	SEDIMENT FILTERS & BARRIERS
	DIVERSIONS		
		SLOPE PROTECTION	MUD & DUST CONTROL

FIGURE 7.11. Erosion Control Measures.

and a planting design scheme which are appropriate to the site and will promote long-term stability. Nonvegetative stabilization measures employ organic and inorganic mulches, gravel, crushed stone, and geofabrics such as meshes, nets, and mats. Diversions are essential in directing runoff away from areas while the vegetative cover is getting established.

Runoff Control

Runoff measures are intended to intercept or control storm water before it has an opportunity to concentrate or before it has reached sufficient volume or velocity to cause damage. Runoff measures include diversions, waterway stabilization, slope protection structures, and outlet protection.

Diversions are channels which direct excess water away from areas where it is not wanted to areas where it can be disposed in an appropriate manner. *Waterways* are natural or constructed channels which provide for the safe disposal of excess water. They may be stabilized vegetatively or with an erosion-resistant lining such as concrete, riprap, or another suitable material. *Check dams,* temporary structures constructed across waterways, may be used to reduce the velocity of concentrated storm water flows, thus reducing the potential for channel erosion. To reduce the risk of erosion, protection is necessary at the outlet of all pipes and paved channels where the flow velocity exceeds the permissible velocity of the receiving channel or area. Structurally lined aprons or other energy-dissipating devices are commonly used.

Sediment Control

Sediment control measures capture sediment on-site before it can be transported downstream or into an existing storm drainage system. Methods of on-site management use sediment basins, silt barriers and fences, filter strips, and storm drain inlet protection. Mud and dust control measures are used to minimize the disturbance and transport of soil by construction vehicles.

The selection of specific erosion control methods and techniques must be suitable to the given site conditions. Most regulatory agencies have developed guidelines for selecting the appropriate measures and specifications for their proper construction and installation. It is the responsibility of the site planner to comply with these regulations.

FIGURE 7.12. Several erosion control measures have been implemented on this construction site. A silt barrier fence has been installed to collect sediment. An outlet pipe in the fence directs runoff toward an existing culvert. Riprap has been used to stabilize the outlet point, and hay bales have been placed across the swale to trap sediment from the outflow before the runoff enters the culvert.

SUMMARY

This chapter has introduced basic storm water management and erosion control techniques and principles. Chapters 8, 9, and 10 discuss methods for calculating storm water runoff volumes, peak discharge rates, flood storage capacity, and system design.

EXERCISES

7.1 Visit several completed site development projects in your region. Identify techniques and measures used to manage storm water runoff.

7.2 Review a copy of the pertinent storm water management regulations (local, county, state, or federal) for the projects visited in Exercise 7.1. Does the installed system, in terms of types of measures used, comply with the governing requirements?

7.3 Visit several site development projects under construction in your region. Identify techniques and measures control erosion and sedimentation caused by storm water runoff.

7.4 Review a copy of the pertinent soil erosion and sedimentation control regulations (local, county, or state) for the projects visited in Exercise 7.3. Does the installed system, in terms of types of measures used, comply with the governing requirements?

8
Determination of Rates and Volumes of Storm Runoff: The Rational and Modified Rational Methods

INTRODUCTION

To design and size storm water management devices, such as grassed swales, drainage pipes, and detention storage ponds, it is first necessary to estimate the rates and volumes of runoff that must be handled. The science of hydrology, which deals with precipitation and runoff, includes a number of models that help in predicting the runoff to be used as input to the design procedures. Many of these are based on precipitation, since for most development sites long-term rainfall records are available. Unlike the deductions of a science, such as hydraulics, which deals with the flow of water in pipes and channels, over weirs, through orifices, and so forth, and in which model studies can be performed in a laboratory, the projections of hydrology cannot be replicated. Every storm or flood event is unique: the durations and intensities of precipitation vary. In addition, the growth stage of vegetation, the activity of soil organisms influencing infiltration rates, and the soil and air temperatures are not likely to be identical for several events. Nevertheless, experience has shown that the forecasts of storm runoff for various probabilities of occurrence are useful for planning and designing storm water management systems and structures. It must be kept in mind, however, that since only estimates are involved, it would be improper to calculate runoff rates and volumes to too great a degree of precision: that is, final results should be appropriately rounded.

RATIONAL METHOD

A frequently used formula for computing the peak rate of runoff from small drainage areas (i.e., less than about 200 acres [ac]) is the *Rational method*. The equation is:

$$q = CiA \qquad (8.1)$$

where: q = peak runoff rate, cubic feet per second (cfs)

C = dimensionless coefficient (between 0 and 1)

i = rainfall intensity, inches per hour (iph) for the design storm frequency and for the time of concentration of the drainage area

A = area of drainage area, acres

The equation is based on the theory that the peak rate of runoff from a small area is equal to the intensity of rainfall multiplied by a coefficient which depends on the characteristics of the drainage area, including land use, soils and slope, and by the size of the area. The extent of the drainage area is determined by connecting the high points and ridge lines on a topographic map or grading plan until a closed system is developed (Fig. 8.1). Drainage areas may vary in size from several hundred square feet for an area drain, or several square miles for a stream, up to thousands of square miles for a large river, in which case it is called a *watershed*. It must also be realized that drainage area boundaries are independent of property lines. It is important to include all parts of a drainage area, even if they are beyond the property line.

The Rational method makes the simplifying assumptions that the rainfall intensity is uniform for the duration of the storm, which must equal at least the time of concentration, and that the precipitation falls on the entire drainage area during that time. These assumptions obviously cannot be applied to large areas.

One acre-inch (ac-in.) per hour of water may be converted to cubic feet per second (cfs) as follows:

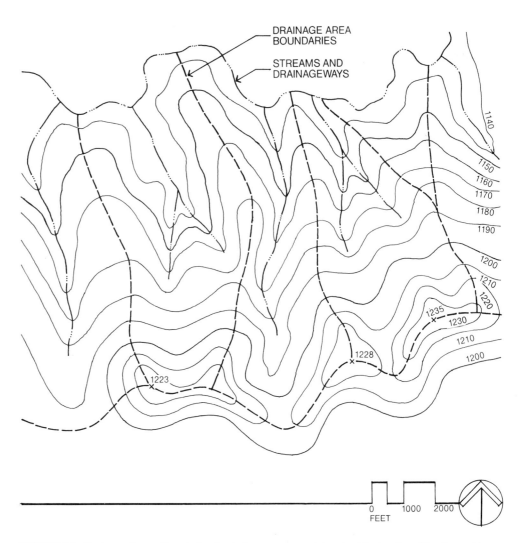

FIGURE 8.1. Drainage Areas. Boundaries for drainage areas are determined by locating ridge lines and high points.

$$1 \text{ ac} \times \frac{1 \text{ in.}}{\text{hr}} \times \frac{43{,}560 \text{ ft}^2}{\text{acre}} \times \frac{1}{12 \text{ in./ft}} \times \frac{1}{3600 \text{ sec/hr}} = 1.008 \text{ cfs}$$

The rate computed by the Rational method is therefore dimensionally incorrect by 0.8%. This is generally ignored, since, as previously stated, hydrology is not an exact science.

The *runoff coefficient, C,* is a value between 0 and 1. Zero represents a completely pervious surface from which there is no runoff; 1 represents a completely impervious and wetted surface from which there is total runoff. Table 8.1 contains suggested C values for a variety of surface conditions. For pervious areas the C value would increase as the soil becomes saturated. Most drainage areas will consist of a variety of surfaces with different C values. Runoff volumes may be computed for each surface separately or an average C value can be computed for the entire drainage area, if the locations of the various land uses are mixed throughout the area.

Rainfall intensity, i, is the rate of rainfall in inches per hour (iph) for the design storm frequency and for the time of concentration of the drainage area. The *storm frequency* is the number of years during which the design storm, or a storm exceeding it, statistically may be expected to occur once. It must be pointed out that the frequency is based on long-term probabilities and that, for example, a 10-yr-frequency storm could conceivably occur several times during a period shorter than 10 yr. This might be compared to throwing a die for which the overall probability of coming up 6 is one out of six throws, but which could come up 6 several times in succession. The *design storm* is a storm with a frequency and duration for which the management system is designed. The selection of a design storm is based on economics, environmental context, and ultimate consequences should the system overflow, and it may be prescribed by applicable regulations. Rainfall intensities for various durations may be obtained from National Weather Service publications and charts similar to Fig. 8.2.

The *time of concentration* (T_c) is the time for water to flow from the hydraulically most remote part of the drainage area to the section under consideration. It is important to realize that this is not necessarily the longest distance, since overland and channel flow time is dependent on slope, surface, and channel characteristics. Assuming a

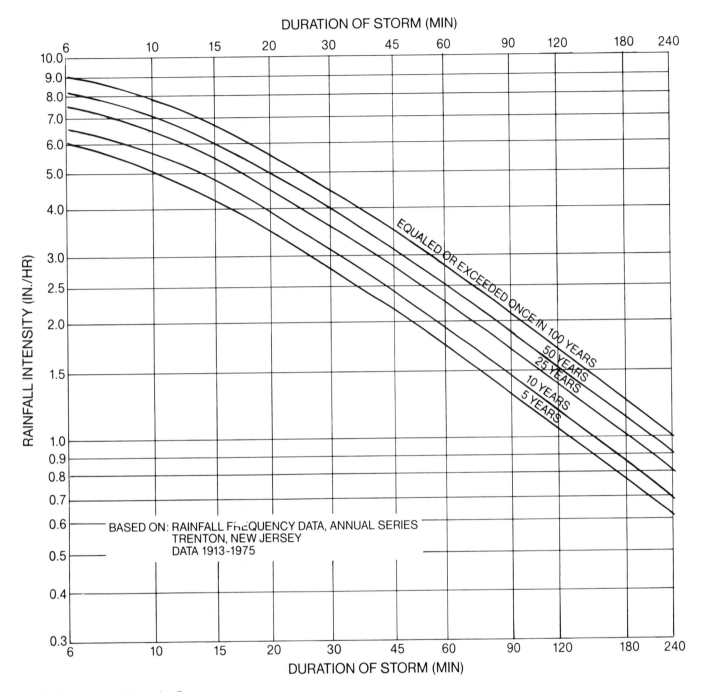

FIGURE 8.2. Rainfall Intensity Curves.

theoretical storm of uniform intensity falling uniformly over the entire drainage area with a duration equal to or exceeding the time of concentration, the maximum runoff is reached at the time of concentration. At this time all parts of the drainage area are contributing simultaneously to the runoff at the section under consideration. Convenient charts which are used to estimate time of concentration are illustrated in Fig. 8.3 for overland flow time, which is demonstrated in Example 8.2 and Fig. 8.4 for channel flow time, which is applied in Example 8.3. The USDA Soil Conservation Service (SCS) method for calculating time of concentration, discussed in the next chapter, may also be used as a basis for determining design storm intensity in the Rational method.

Example 8.1

The drainage area for a grassed waterway is an 8.2-ac site for a small industrial plant. As shown in Fig. 8.5, several different surfaces are dispersed throughout the site. These surfaces include 4.4 ac of relatively flat lawn with a silt loam soil, 2.8 ac of pavement, and 1.0 ac of roof surface. To simplify this example the rainfall intensity is assumed to be 2 iph. Calculate the peak rate of runoff for this drainage area.

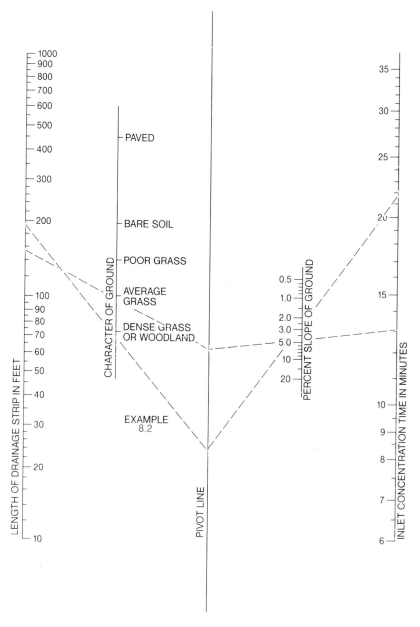

FIGURE 8.3. Nomograph for Overland Flow Time.

Solution 1

$A = 4.4$ ac lawn, 2.8 ac pavement, 1.0 ac roof

$i = 2$ iph

From Table 8.1:

$C_{\text{flat lawn}} = 0.30$

$C_{\text{pavement}} = 0.85$

$C_{\text{roof}} = 0.95$

Using the equation $q = CiA$, substitute the known values.

$q_{\text{lawn}} = 0.30 \times 2.0 \times 4.4$

$q_{\text{lawn}} = 2.64 \approx 2.6$ cfs

$q_{\text{pavt.}} = 0.85 \times 2.0 \times 2.8$

$q_{\text{pavt.}} = 4.76 \approx 4.8$ cfs

$q_{\text{roof}} = 0.95 \times 2.0 \times 1.0$

$q_{\text{roof}} = 1.9$ cfs

$q_{\text{total}} = 2.6 + 4.8 + 1.9 = 9.3$ cfs

Solution 2

$A_{\text{total}} = 4.4 + 2.8 + 1.0 = 8.2$ ac

$i = 2$ iph

$C_{\text{avg.}} = \dfrac{(4.4 \times 0.30) + (2.8 \times 0.85) + (1.0 \times 0.95)}{8.2}$

$C_{\text{avg.}} = 0.567 \approx 0.57$

Again substitute the known values into the equation $q = CiA$:

$q_{\text{total}} = 0.57 \times 2.0 \times 8.2$

$q_{\text{total}} \approx 9.3$ cfs

Note that the total q value is the same whether it is computed separately or by using an average C value. However, for this example the average C value provides a more descriptive representation of the actual runoff conditions for the drainage area.

Example 8.2

The 1.5-ac drainage area in New Jersey illustrated in Fig. 8.6 consists of 0.14 ac of pavement, 0.20 ac of woodland, and 1.16 ac of lawn. The soil is a silt loam and the slopes range up to 5%. Calculate the peak rate of runoff leaving the site at the southwest corner. A 10-yr design storm is to be used.

Solution. The first step in this problem is to determine the time of concentration in order to establish the rainfall

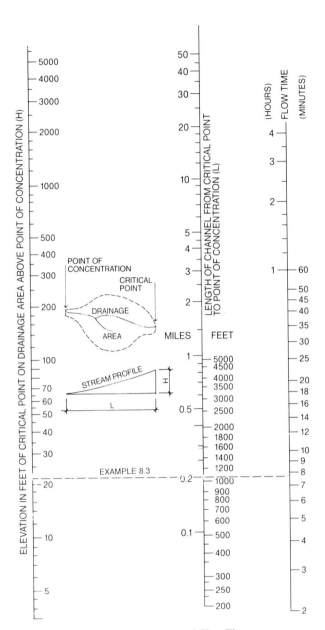

FIGURE 8.4. Nomograph for Channel Flow Time.

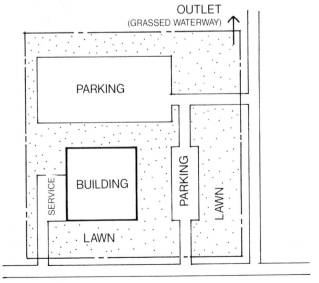

FIGURE 8.5. Schematic Site Plan for Example 8.1.

TABLE 8.1. Recommended Runoff Coefficients *(C)*

Urban areas[a]	
Downtown Business	0.70–0.95
Neighborhood Business	0.50–0.70
Single-family Residential	0.30–0.50
Detached Multi-Unit Residential	0.40–0.60
Attached Multi-Unit Residential	0.60–0.75
Suburban Residential	0.25–0.40
Apartment	0.50–0.70
Light Industry	0.50–0.80
Heavy Industry	0.60–0.90
Parks, Cemeteries	0.10–0.25
Playgrounds	0.20–0.35
Railroad Yards	0.20–0.35
Unimproved	0.10–0.30
Urban surfaces	
Roofs	0.80–0.95
Asphalt and Concrete Pavements	0.75–0.95
Gravel	0.35–0.70

	Soil Texture		
Rural and suburban areas[b]	Sandy loam	Clay and silt loam	Clay
Woodland			
flat (0–5% slope)	0.10	0.30	0.40
rolling (5–10% slope)	0.25	0.35	0.50
hilly (10–30% slope)	0.30	0.50	0.60
Pasture and Lawns			
flat	0.10	0.30	0.40
rolling	0.16	0.36	0.55
hilly	0.22	0.42	0.60
Cultivated or No Plant Cover			
flat	0.30	0.50	0.60
rolling	0.40	0.60	0.70
hilly	0.52	0.72	0.82

[a] From American Iron and Steel Institute (1980)
[b] From Schwab et al. (1971)

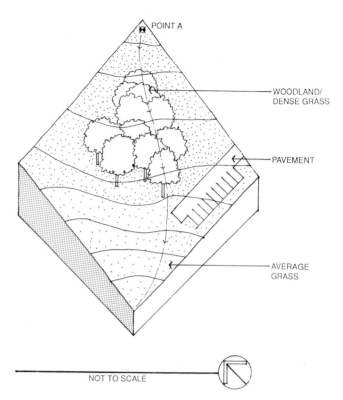

FIGURE 8.6. Schematic Site Plan for Example 8.2.

intensity value, i. Additional information is required to compute the overland flow time. From visual inspection it is determined that point A is the hydraulically most remote point on the site. The *distance* that water must travel to reach the outlet point from point A, the *slope* of its path of travel, and the *surface character* of the path must also be determined. This is summarized here for this problem.

Length	Character	Slope	Time of Concentration
200 ft	Woodland/ dense grass	2%	22 min
160 ft	Average grass	4%	13 min
			35 min total

The time of concentration is determined from the chart in Fig. 8.3. First the travel distance is located along the length line, then the surface character is located on the character of ground line. A straight line is drawn through these two points until it intersects the pivot line. Next the slope of the path is located on the percent slope line and a straight line is drawn from the point on the pivot line through the point on the slope line until it intersects the inlet concentration time. This point of intersection represents the time of concentration in minutes. No additional time needs to be computed, since there is no channel flow involved in this example.

The intensity of the design storm can then be obtained from the chart in Fig. 8.2. Locate the 35-min storm duration along the horizontal axis of the chart and extend a vertical line until it intersects the 10-yr frequency curve. From the point on the curve extend a horizontal line to the left until it intersects the vertical axis. The value obtained, about 2.8 iph, is the rainfall intensity for the storm duration. Note that the scales are logarithmic. The chart in Fig. 8.2 applies to New Jersey. The appropriate charts or similar information for other states may be obtained from the state department of conservation, water resources, or other similar agency, or from the National Weather Service.

The next step is to select an appropriate runoff coefficient from Table 8.1 for the various surfaces.

$C_{pavement} = 0.90$

$C_{woodland} = 0.30$

$C_{lawn} = 0.30$

The last step is to substitute all the known values into the equation $q = CiA$.

$q_{total} = (0.90 \times 2.8 \times 0.14) + (0.30 \times 2.8 \times 0.20) + (0.30 \times 2.8 \times 1.16)$

$$q_{total} = 2.8 \times (0.90 \times 0.14 + 0.30 \times 0.20 + 0.30 \times 1.16)$$

$$q_{total} = 2.8 \times 0.53 = 1.484 \approx 1.5 \text{ cfs}$$

The following example demonstrates the determination of the time of concentration when overland flow and stream flow are involved.

Example 8.3

Figure 8.7 shows a 95-ac drainage area which is completely wooded with a stream flowing through the center. The soil survey report indicates a silt loam. The topographic map shows that the average drainage area slope is 6%, while the stream slope is 2%. A road culvert is to be installed at the outlet from this drainage area and must be designed for a 50-yr storm frequency. What is the peak rate of runoff for which the culvert must be designed?

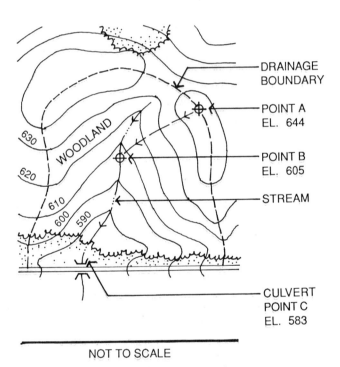

FIGURE 8.7. Schematic Site Plan for Example 8.3.

Solution. An examination of the topographic map shows that A is the hydraulically most remote point. From A to B the water flows overland for 970 ft along a 4% slope. (The *average* drainage area slope is 6%.) From B to C there is 1100 ft of stream flow with a 2% slope. Using Fig. 8.3 as before, the overland flow here is determined as 35 min. To obtain the stream flow time Fig. 8.4 must be used. To calculate H, which is the elevation of the critical point above the point of concentration, that is, the elevation of point B above point C, distance B–C is multiplied by the slope:

$$H = 1100 \times 0.02 = 22 \text{ ft}$$

A straight line is drawn from 22 ft on the H scale on the left through 1100 ft on the L scale to the T scale on the right, which yields about 8 min. (Note the rapidity of the stream flow compared to overland flow.) Then the total time of concentration is overland flow time + stream flow time, or $35 + 8 = 43$ min. The intensity for a 50-year 43-min storm is about 3.3 iph from Fig. 8.2. The C value for woodland on silt loam with an average slope of 6% is 0.35 from Table 8.1. The design peak rate of runoff is thus:

$$q = CiA$$
$$= 0.35 \times 3.3 \times 95$$
$$\approx 110 \text{ cfs}$$

Therefore, the culvert must be sized to accommodate 110 cfs.

As demonstrated by the following examples, common sense should be applied when determining time of concentration, particularly with regard to the location of different surface types (C values) on a site.

Example 8.4A

Figure 8.8(a) schematically illustrates a drainage area which consists of 3 ac of pavement ($C = 0.80$) and 3 ac of lawn ($C = 0.30$). As contrasted with Example 8.1, where the different surfaces were mixed throughout the site, these two surfaces are concentrated and segregated. Determine the peak rate of runoff at the point of discharge from the site for a 10-yr storm frequency. Assume a 6-min travel time over the pavement and a 30-min travel time over the lawn. The runoff from the pavement flows over the lawn area to the discharge point.

Solution. Since it takes the hydraulically most remote drop of water 6 min to flow over the pavement and an additional 30 min to flow over the lawn before reaching the discharge point, the total time of concentration is 36 min $(6 + 30)$. Entering the chart in Fig. 8.2 with a storm duration of 36 min, a rainfall intensity of approximately 2.8 iph is determined from the 10-yr storm frequency curve. From the Rational formula, the peak rate of runoff is calculated as follows:

$$q = CiA$$
$$q_{lawn} = 0.30 \times 2.8 \times 3.0$$
$$q_{lawn} = 2.52 \approx 2.5 \text{ cfs}$$
$$q_{pavt.} = 0.80 \times 2.8 \times 3.0$$
$$q_{pavt.} = 6.72 \approx 6.7 \text{ cfs}$$
$$q_{total} = 2.5 + 6.7 = 9.2 \text{ cfs}$$

Example 8.4B

All the conditions in this example, including travel times, are the same as in Example 8.4A except the relationship of pavement to lawn has been reversed. Now the runoff

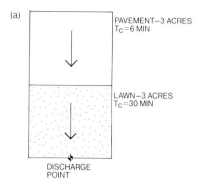

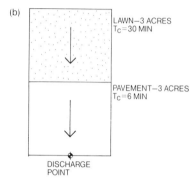

FIGURE 8.8. Schematic Plans. a. Example 8.4a. b. Example 8.4b.

from the lawn flows over the pavement to the discharge point as illustrated in Fig. 8.8b. Again calculate the peak rate of runoff from the site using a 10-yr storm frequency.

Solution. In the previous example the lawn attenuated the velocity of flow from the pavement, since it was located between the paved area and the point of discharge. However, in this example the flow from the pavement is not attenuated and its runoff concentrates at the discharge point in 6 min. Therefore, the storm duration for determining the rainfall intensity is 6 min for the pavement. Again, referring to Fig. 8.2, the rainfall intensity is 6.6 iph on the 10-yr storm frequency curve. Thus, the peak rate of runoff from the pavement is:

$q_{pavt.} = 0.80 \times 6.6 \times 3.0$

$q_{pavt.} = 15.84 \approx 15.8$ cfs

The runoff rate from the pavement alone is considerably greater than the runoff rate for the entire drainage area in Example 8.4A.

This example demonstrates that great care must be taken to arrive at a realistic estimate of runoff rates. Average C values should not be used unless the various land uses are distributed uniformly throughout a drainage area. For example, if a 40-ac housing development with streets, gutters, and storm sewers were to be constructed in a 200-ac wooded drainage area, the runoff from the dissimilar land uses should be determined separately and then added. The next section will include procedures for accomplishing this.

MODIFIED RATIONAL METHOD

The Rational method uses runoff coefficients (C) that are constant for all frequencies and all storms and yields only peak runoff rates, using rainfall intensities for a specific storm frequency and duration. Normally, the times of concentration on which these intensities are based are relatively short. It has been recognized that for the less frequent storm events, that is, those which may occur with frequencies of 25 to 100 yr, the short times of concentration do not adequately represent the total duration of rainfall with its resulting runoff. The durations of storm events used for the Rational method, equal to the time of concentration, are usually parts of larger storm systems with periods of rainfall preceding and extending beyond the time of concentration. Much of the soil's ability to absorb storm runoff has been exhausted by "antecedent precipitation."

The Modified Rational method (MRM) recommends that an *antecedent precipitation factor* (C_A) be included in the Rational formula, so that it becomes:

$$q = CC_A iA \qquad (8.2)$$

Recommended values for C_A are listed in Table 8.2. The product of $C \times C_A$ must never exceed 1, since otherwise the computed runoff rate would be greater than the intensity of the design storm, which is incongruous. The American Public Works Association suggests that the use of the MRM for the determination of required storage volumes be limited to small drainage areas, that is, those less than 20 ac.

Although peak runoff rates are sufficient for sizing structures, such as culverts, waterways, and pipes, the design of storage ponds and reservoirs requires determination of inflow and outflow volumes, obtained from hydrographs. A *hydrograph* is a plot of flow rate (q) versus time (T) (Fig. 8.9) or the tabular representation of such plots. Hydrographs are labeled by the duration (DUR) of the *rainfall events* producing them, not by the duration of the *runoff*. Thus, a 30-min storm produces a 30-min hydrograph, regardless of the duration of the runoff.

Some situations also require the determination of flow rates at any time during the runoff process which can be obtained from hydrographs. The MRM also permits the determination of runoff for storm durations shorter or longer than the time of concentration.

TABLE 8.2. Recommended Antecedent Precipitation Factors

Frequency (years)	C_A
2 to 10	1.0
25	1.1
50	1.2
100	1.25

[a] From American Public Works Association (1974)

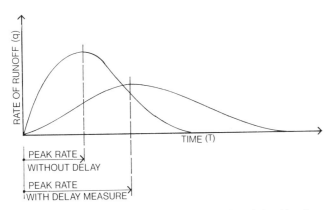

FIGURE 8.9. Hydrographs. Hydrographs plot the relationship of runoff rate to time. When land development takes place and delay measures such as retention and detention basins are not used, the peak occurs earlier and at a higher rate than if delay measures are used as qualitatively illustrated in the graph.

Three types of simplified hydrographs can be developed by the MRM. For all three types the intensity used to calculate the maximum runoff rate (q_{max}) is that for the duration and frequency of the storm event.

Type A Hydrograph

The duration of the storm event is equal to the time of concentration as shown in Fig. 8.10.

$$q_p = CC_A iA \qquad (8.3)$$

The peak rate of runoff (q_p) is reached at the time of concentration (T_c). The time of recession (T_{rec}) equals the time of concentration. This hydrograph represents the maximum potential runoff rate from the drainage area for the specified frequency.

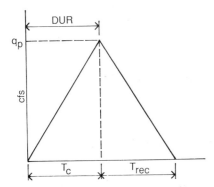

FIGURE 8.10. Type A Hydrograph.

Type B Hydrograph

As illustrated in Fig. 8.11, the duration of the storm is greater than the time of concentration.

$$q_{max} = CC_A iA \qquad (8.4)$$

The maximum flow rate (q_{max}) is reached at the time of concentration but is smaller than that of the type A hydrograph, since the intensity for the longer duration is less. The runoff rate remains constant to the end of the storm duration. The time of recession equals the time of concentration.

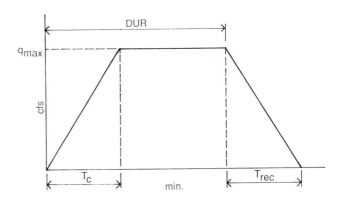

FIGURE 8.11. Type B Hydrograph.

Type C Hydrograph

The duration of the storm, as depicted in Fig. 8.12, is less than the time of concentration.

$$q_{max} = CC_A iA(\mathrm{DUR}/T_c) \qquad (8.5)$$

The maximum runoff rate (q_{max}) is reached at the end of the storm duration. It remains constant to the end of the time of concentration. The time of recession equals the duration of the storm. The intensity of the storm is greater than that used for the type A hydrograph. However, since the drainage area has not reached its time of concentration and maximum potential runoff rate, the reduction factor (DUR/T_c) is included in the equation.

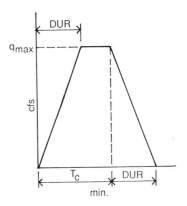

FIGURE 8.12. Type C Hydrograph.

Example 8.5

Develop 10-, 25-, and 50-min hydrographs for a 15-ac drainage area, having a C value of 0.3 and a time of concentration of 25 min. A 100-yr frequency is desired.

Solution: 10-Minute Hydrograph. Since the storm duration is less than the time of concentration, this is a type C hydrograph and the equation used is:

$q_{max} = CC_A iA(DUR/T_c)$

$C_A = 1.25$ (from Table 8.2)

$i = 7.9$ iph (approx.) (from Fig. 8.2)

$q_{max} = 0.3 \times 1.25 \times 7.9 \times 15 \times \dfrac{10}{25}$

$q_{max} \approx 17.8$ cfs

Note that although the duration of the storm event is only 10 min, the duration of runoff is 35 min (T_c + DUR).

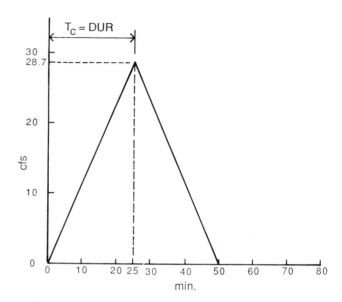

FIGURE 8.14. 25-Minute Hydrograph for Example 8.5.

$q_{max} = CC_A iA$

$C_A = 1.25$ (from Table 8.2)

$i = 3.4$ iph (approx.) (from Fig. 8.2)

$q_{max} = 0.3 \times 1.25 \times 3.4 \times 15$

$q_{max} \approx 19.1$ cfs

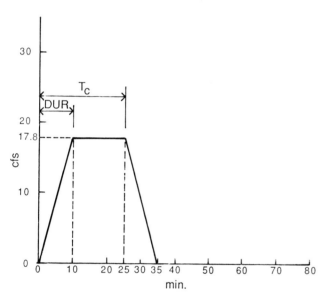

FIGURE 8.13. 10-Minute Hydrograph for Example 8.5.

Solution: 25-Minute Hydrograph. Since the duration of the storm event equals the time of concentration, this is a type A hydrograph. The equation used is:

$q_p = CC_A iA$

$C_A = 1.25$ (from Table 8.2)

$i = 5.1$ iph (approx.) (from Fig. 8.2)

$q_p = 0.3 \times 1.25 \times 5.1 \times 15$

$q_p \approx 28.7$ cfs

Solution: 50-Minute Hydrograph. Since the duration of the storm event is greater than the time of concentration, this is a type B hydrograph.

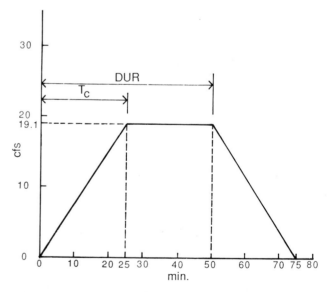

FIGURE 8.15. 50-Minute Hydrograph for Example 8.5.

Hydrologic theory permits the combination of runoff hydrographs emanating from several subareas and then joining in one flow path. The next example will demonstrate this procedure.

Example 8.6

In Example 8.4B the peak rate of runoff was calculated for the pavement only. By means of separate hydrographs, determine the total runoff from the lawn and pavement.

Solution. The runoff hydrograph from the pavement is type A, since the storm duration equals the time of concentration of 6 min. Listing the runoff by 6-min intervals, the hydrograph can be represented as follows:

Time, min	Flow Rate, cfs
0	0.0
6	15.8
12	0.0

(See computations in the solution to Example 8.4B.)

The hydrograph for a 6-min duration and a 36-min time of concentration for the lawn area is type C, since the duration is shorter than the time of concentration.

$q_{max} = CC_A iA \,(\text{DUR}/T_c)$

$C_A = 1.0$ (for a 10-yr frequency)

$i = 6.6$ iph

$q_{max} = 0.3 \times 1.0 \times 6.6 \times 3 \times \left(\dfrac{6}{36}\right)$

$q_{max} = 0.99 \approx 1.0$ cfs

The hydrograph for the lawn is tabulated as follows:

Time, min	Flow Rate, cfs
0	0.0
6	1.0
12	1.0
18	1.0
24	1.0
30	1.0
36	1.0
42	0.0

The two hydrographs may be combined as follows:

Time, min	Flow rate from pavement, cfs	Flow rate from lawn, cfs	Total runoff, cfs
0	0.0	0.0	0.0
6	15.8	1.0	16.8
12	0.0	1.0	1.0
18	0.0	1.0	1.0
24	0.0	1.0	1.0
30	0.0	1.0	1.0
36	0.0	1.0	1.0
42	0.0	0.0	0.0

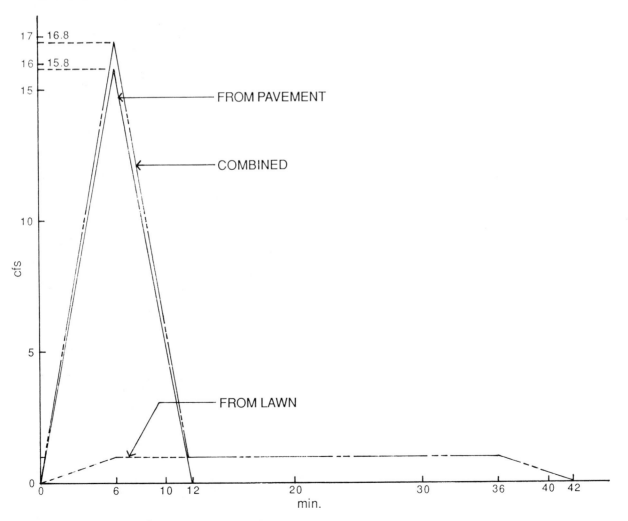

FIGURE 8.16. Combined Hydrograph for Example 8.6.

112   Determination of Rates and Volumes of Storm Runoff: The Rational and Modified Rational Methods

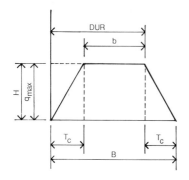
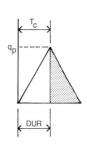
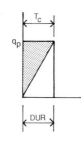

(a)

a. Type A hydrograph.
By geometry, the area of a triangle is $A = \frac{1}{2} B \times H$.
For hydrographs: if A = volume
$$B = 2 \times \text{duration}$$
$$H = q_p$$
Then Vol $= \frac{1}{2} \times (2 \times \text{DUR}) \times q_p$
$\qquad = \text{DUR} \times q_p.$

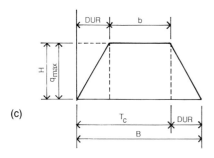
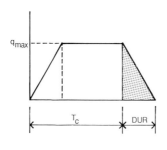
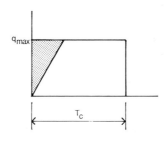

(b)

b. Type B hydrograph.
By geometry, the area of a trapezoid is $A = [(B + b)/2] \times H$.
For hydrographs: if A = volume
$$B = \text{duration} + T_c$$
$$b = \text{duration} - T_c$$
$$H = q_{max}$$
Then Vol $= [(\text{DUR} + T_c + \text{DUR} - T_c)/2] \times q_{max}$
$\qquad = [(2 \times \text{DUR})/2] \times q_{max}$
$\qquad = \text{DUR} \times q_{max}$

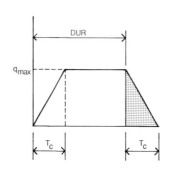
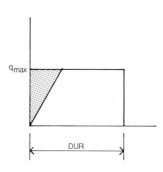

(c)

c. Type C hydrograph.
Geometry similar to type B hydrograph.
If A = volume
$$B = T_c + \text{DUR}$$
$$b = T_c - \text{DUR}$$
$$H = q_{max}$$
Then Vol $= [(T_c + \text{DUR} + T_c - \text{DUR})/2] \times q_{max}$
$\qquad = [(2 \times T_c)/2] \times q_{max}$
$\qquad = T_c \times q_{max}$

FIGURE 8.17. Volume Diagrams.

Although the runoff from the lawn is ignored in Example 8.4B, the combined hydrograph shows that the lawn area is contributing 1 cfs at the 6-min T_c for the pavement. Therefore, the peak runoff rate is 16.8 cfs, not 15.8 as previously computed (Fig. 8.16). Including the lawn in this example does not change the outcome significantly; however, considerable differences may result for larger drainage areas.

VOLUMES OF RUNOFF, STORAGE, AND RELEASE

The area under a hydrograph represents the volume of runoff (Vol). If the q values on the y axis are in cubic feet per second (cfs) and the T values on the x axis are converted into seconds, the volume can be computed as

$$\text{Vol} = \text{cu ft/sec} \times \text{sec} = \text{cu ft (cfs} \times \text{sec} = \text{ft}^3) \quad (8.6)$$

Since the MRM specifies for each type of hydrograph that the time of recession must equal the time of rise,[6] the volumes for type A and B hydrographs are the products of the *duration* in seconds times the maximum flow rate in cfs. For type C hydrographs the volume is the product of the *time of concentration* in seconds times the maximum flow rate in cfs. This may be seen conceptually through the diagrams in Fig. 8.17.

Example 8.7

Compute the runoff volumes for the 25- and 50-min storm durations for the drainage area of Example 8.5.

Solution: 25-Minute Duration. The first step is to convert the duration from minutes to seconds (25 min × 60 sec/min = 1500 sec). The peak runoff rate for the 25-min storm is 28.7 cfs from the solution for the 25-min hydrograph. The volume (Vol) for the 25-min storm is computed as follows:

Vol = 28.7 cfs × 1500 sec = 43,050 ft³

Solution: 50-Minute Duration. The duration in seconds is 50 min × 60 sec/min = 3000 sec. The maximum runoff rate for the 50-min storm is 19.1 cfs from the solution for the 50-min hydrograph. The volume (Vol) for the 50-min storm is computed as follows:

Vol = 19.1 cfs × 3000 sec = 57,300 ft³

This shows that short and intense storms (with corresponding high i values and peak runoff rates) do not produce the greatest runoff volumes.

REQUIRED STORAGE FOR DETENTION OR RETENTION PONDS BY THE MODIFIED RATIONAL METHOD

The most important use of the MRM is the determination of required storage volumes for detention or retention ponds. The procedure is to subtract the volume of outflow from the basin from the inflow volume to the basin for the same duration. The difference is the required storage.

Usually, the initial duration used is the time of concentration. However, as previously shown in Example 8.7, short, intense storm events do not necessarily yield the greatest volumes. Therefore, the duration is then increased from the time of concentration by equal time intervals until the maximum required storage is reached. The rate of outflow is generally prescribed by governmental regulations. Most often the maximum postdevelopment outflow for specified frequencies cannot exceed the predevelopment rate for the same frequencies.

Example 8.8

Determine the required storm water storage if the 15-ac drainage area of Example 8.5 is developed so that the average C value is increased to 0.75 and the time of concentration is reduced to 15 min. The local ordinance requires that the postdevelopment 100-yr outflow from the area cannot exceed the 28.7-cfs predevelopment rate previously computed for the 25-min T_c.

Solution. Starting with the time of concentration of 15 min, i for a 100-yr frequency and a 15-min duration is 6.8 iph from Fig. 8.2. C_A is 1.25 from Table 8.2. Using the type A hydrograph the equation is:

$q_p = CC_A iA$

$q_p = 0.75 \times 1.25 \times 6.8 \times 15$

$q_p = 95.6$ cfs inflow rate to pond

Vol$_i$ = 95.6 cfs × 15 min × 60 sec/min = 86,040 ft³ inflow volume

Vol$_o$ = 28.7 cfs × 15 min × 60 sec/min = 25,830 ft³ outflow volume

The storage volume required for a 100-yr 15-min storm event is:

Vol$_{stor}$ = 86,040 ft³ inflow − 25,830 ft³ outflow = 60,210 ft³

Using a 5-min time increment, the required storage for a 20-min storm is determined next. The intensity, i, for a

---

[6]This is obviously an oversimplification and there are other systems that use different proportions. However, for the formulas used here, these configurations, as recommended by the American Public Works Association, must be used.

TABLE 8.3. Required Storage at 5-Minute Intervals for Example 8.8

Storm Frequency, yr	Duration, min	Intensity, iph[a]	C_a[b]	Maximum Inflow, cfs[c]	Maximum Outflow, cfs[d]	Inflow Volume, cu ft[e]	Outflow Volume, cu ft[f]	Required Storage, cu ft[g]
100	15	6.8	1.25	95.6	28.7	86040	25830	60210
100	20	5.7	1.25	80.2	28.7	96240	34440	61800
100	25	5.1	1.25	71.7	28.7	107550	43050	64500
100	30	4.6	1.25	64.7	28.7	116460	51660	*64800
100	35	4.2	1.25	59.1	28.7	124110	60270	63840

[a] From Fig. 8.2
[b] From Table 8.2
[c] From $q = CC_A iA$ for postdevelopment conditions
[d] Allowable outflow rate for predevelopment conditions
[e] From duration (sec) × inflow rate (cfs)
[f] From duration (sec) × outflow rate (cfs)
[g] Inflow volume − outflow volume
* Maximum, or critical, storage volume

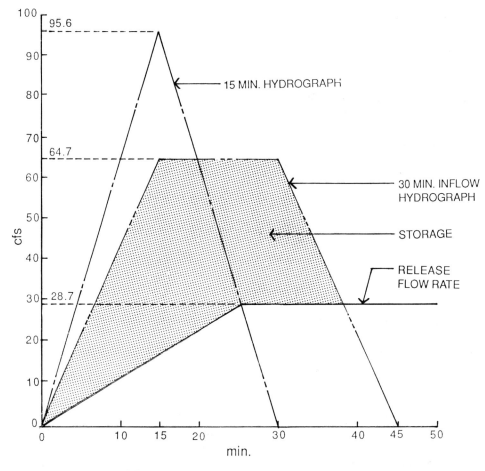

FIGURE 8.18. Inflow and Outflow Hydrographs for Example 8.8.

100-yr frequency and a 20-min duration is 5.7 iph from Fig. 8.2.

$q_{max} = CC_A iA$

$q_{max} = 0.75 \times 1.25 \times 5.7 \times 15$

$q_{max} = 80.2$ cfs inflow rate to pond

$\text{Vol}_i = 80.2 \text{ cfs} \times 20 \text{ min.} \times 60 \text{ sec/min} = 96,240 \text{ ft}^3$ inflow volume

$\text{Vol}_o = 28.7 \text{ cfs} \times 20 \text{ min} \times 60 \text{ sec/min} = 34,440 \text{ ft}^3$ outflow volume

The storage volume required for a 100-yr 20-min storm event is:

Vol_{stor} = 96,240 ft³ inflow − 34,440 ft³ outflow = 61,800 ft³

The procedure is continued with 5-min intervals. The results are presented in Table 8.3.

In this case, the maximum or "critical" storage requirement for a 100-yr event results from a storm duration of 30 min. The outlet structure for the pond should discharge 28.7 cfs when the pond contains 64,800 ft³. It must be emphasized that the procedure yields a volume which is only an estimate, possibly sufficient for a small pond in a situation where failure would not cause serious damage and liability. Where these conditions are not met, the results produce a starting point for design and the actual required volume must be determined by a detailed analysis of the proposed outlet structure and the actual pond dimensions, which is called *reservoir routing*. A discussion of this topic is beyond the scope of this text. Storm water management regulations may require the routing of the runoff from additional storm frequencies.

The inflow and outflow hydrographs for this example are illustrated in Fig. 8.18. The maximum release flow does not take place immediately. The MRM specifies that it occurs at the point where the release rate intersects the recession portion of a hydrograph having a duration equal to the time of concentration (type A hydrograph): that is, 15-min for this example.

SUMMARY

This chapter has described the Rational method for determining storm water surface runoff rates. It has introduced the concept of time of concentration for a drainage area and explained the Modified Rational method of developing simplified hydrographs and determining volumes of runoff and required storm water storage. The following chapter will deal with the Soil Conservation Service methods.

EXERCISES

8.1 A drainage area consists of 8 ac of lawn ($C = 0.36$) and 2 ac of pavement and roofs ($C = 0.90$). By the Rational method determine the peak rate of runoff for a 10-yr storm frequency, to be used for the design of a grassed swale, if the time of concentration is 45 min.

8.2 (a) Examination of the topographic and land use maps of a drainage area indicates that the flow from the hydraulically most remote point to the point of concentration proceeds as follows:

300 ft of woodland with a 4% slope
200 ft of dense grass with a 3% slope
500 ft of average grass with a 2% slope
1600 ft of stream flow with a 1% slope

Find the total time of concentration for this area.

(b) If the area of part (a) contains 125 ac of which 35% is in woodland, 64% is in pasture, and 1% is in gravel roads, determine the peak storm runoff rate for a 25-yr frequency. All surface slopes of the area are less than 5%.

8.3 (a) By the MRM, determine the 100-yr peak flow rate for a 9.5-ac drainage area with the following characteristics:

Average land slope 1%
Average grass ground cover
Length of overland flow path 900 ft

(b) Develop and sketch hydrographs for the area of part (a) for the following durations:

½ T_c
T_c
2 × T_c

8.4 (a) The area of Exercise 8.3 is to be developed for light industry, and the time of concentration will be reduced to 20 min. Determine the 100-yr peak runoff rate by the MRM.

(b) What will be the required maximum (critical) storage volume for the area in (a) if the postdevelopment 100-yr outflow rate cannot exceed the predevelopment rate? What will be the duration of the "critical" storm?

9
Soil Conservation Service Methods of Estimating Runoff Rates, Volumes, and Required Detention Storage

INTRODUCTION

In addition to the United States Department of Agriculture Soil Conservation Service (SCS), many other government agencies and private consultants have adopted the use of SCS hydrologic procedures. These are based on runoff, in inches, resulting from rainfall, in inches, for a specific drainage area and land use, and for a specified storm duration. Normally this duration is far greater than the time of concentration, in contrast to the Rational method, and therefore the rainfall intensity is not constant for the design storm event. However, the assumption that rain falls equally on the entire drainage area still applies.

The inches of precipitation are transformed into inches of runoff by means of a runoff curve number (CN), based on soils, land use, impervious areas, interception by vegetation and structures, and temporary surface storage. Figure 9.1 illustrates the relationship between curve number and direct runoff. Hydrographs are developed by means of unit hydrograph theory and the modified attenuation-kinematic routing method. A discussion of these two procedures is not within the scope of this text and this chapter will be limited to the use of SCS Technical Release 55, *Urban Hydrology for Small Watersheds* (TR55).

TR55 is the most commonly used guide to the application of SCS hydrology. It is a simplified condensation of SCS hydrologic procedures, suitable for drainage areas with times of concentration (T_c) of up to 2 hr for the Tabular Hydrograph method and up to 10 hr for the Graphical Peak Discharge method. The publication contains numerous tables and graphs of which only samples are included in this text. For application to actual projects the reader is urged to obtain the source document.

RAINFALL

Four regional rainfall patterns are used by the SCS. Their approximate geographical boundaries are shown in Fig. 9.2. The cumulative proportions of total rainfall depth for all frequencies over a 24-hr period are illustrated in Fig. 9.3 for all four distributions. An "intermediate storm" pattern has been used; there is a period of high intensity (a large rainfall amount in a relatively short period), preceded and followed by more gradual accumulation. For type II and type III storms, approximately one-half of the 24-hr precipitation occurs in 1 hr at about the midpoint.

Although SCS hydrology can be applied to storm events of various lengths, a 24-hr duration *must* be used for all applications involving TR55. Twenty-four-hour rainfall depths for frequencies from 2 to 100 yr for the eastern and central United States can be interpolated from Figs. 9.4 to 9.7. Values for areas not included in the maps can be obtained from SCS state conservation engineers for those areas.

PROCEDURES OF TR55

Figure 9.8 is a flow chart for determining appropriate procedures to be used for various purposes. For example, to estimate peak flow from an area with homogeneous topographic, soil, and land use conditions, the Graphical Discharge method could be used. However, if there are several subareas in a watershed, each with different characteristics, the Tabular Hydrograph method would have to be used to arrive at the total runoff rate. TR55 deals with surface flow over land and in open channels only. In drainage areas having storm sewers, flow paths and times for large events should be carefully investigated. Most of the

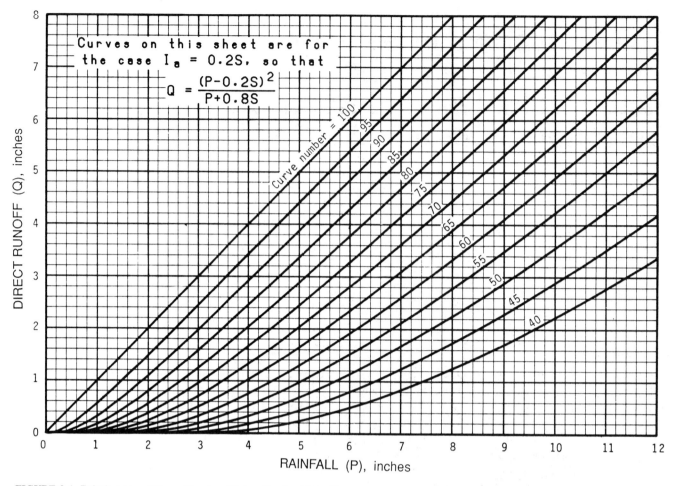

FIGURE 9.1. Relationship of Curve Number (CN) to Depth of Runoff.

FIGURE 9.2. SCS Rainfall Distribution Patterns.

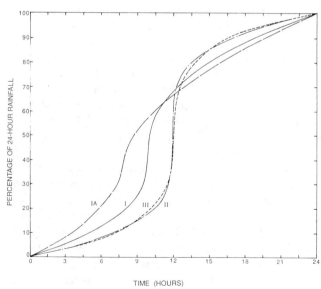

FIGURE 9.3. SCS 24-Hour Rainfall Distributions.

Procedures of TR55   119

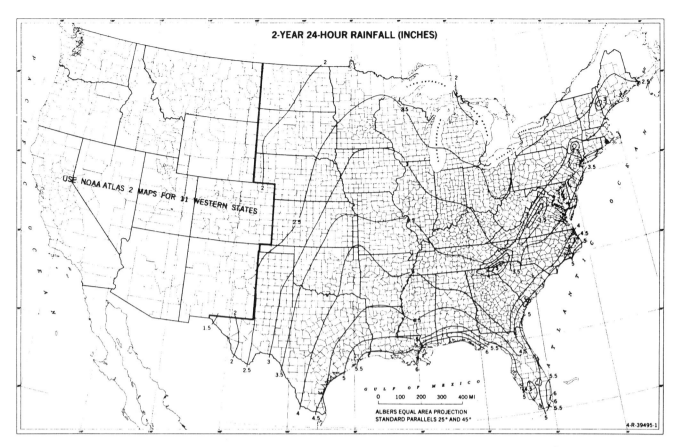

FIGURE 9.4. 2-Yr, 24-Hour Rainfall.

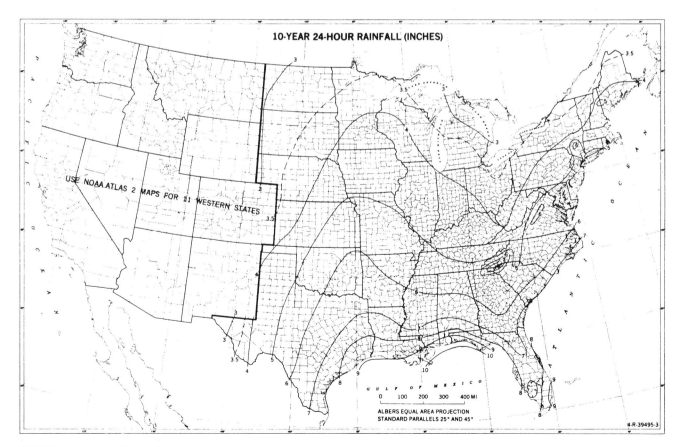

FIGURE 9.5. 10-Yr, 24-Hour Rainfall.

120 Soil Conservation Service Methods of Estimating Runoff Rates, Volumes, and Required Detention Storage

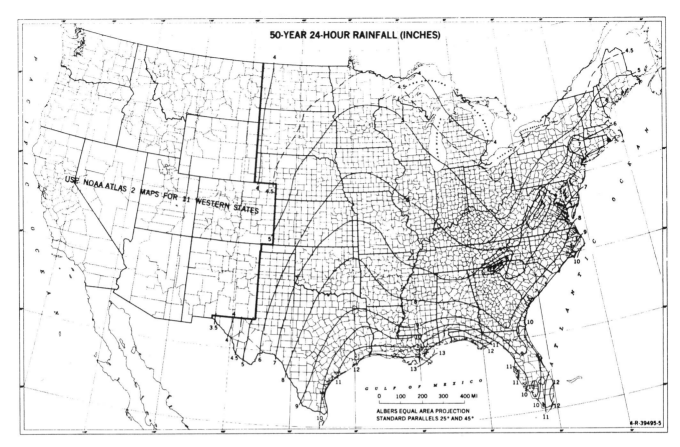

FIGURE 9.6. 50-Yr, 24-Hour Rainfall.

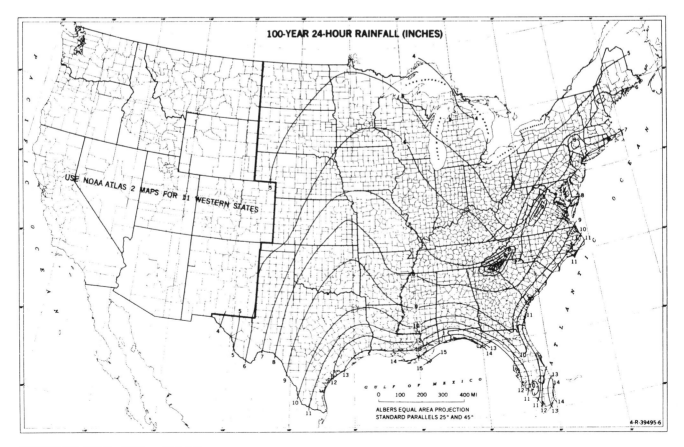

FIGURE 9.7. 100-Yr, 24-Hour Rainfall.

Procedures of TR55   121

$$T_t = \frac{L}{V} \qquad (9.1)$$

where: T_t = travel time, sec

L = flow length, ft

V = flow velocity, ft/sec (fps)

(Divide T_t by 3600 sec/hr to convert to hours of travel time.)

The SCS separates flow into three distinct processes: sheet flow, shallow concentrated flow, and open channel flow. *Sheet flow* is flow over plane, sloped surfaces and normally takes place at the upper end of a flow path. Its maximum permitted length is 300 ft; however, it may be less and its actual length is best determined by field inspection and observation. Manning's kinematic equation is used to determine travel time for sheet flow:

$$T_t = \frac{0.007 \, (nL)^{0.8}}{(P_2)^{0.5} S^{0.4}} \qquad (9.2)$$

where: T_t = travel time, hrs

n = Manning's roughness coefficient for sheet flow (Table 9.1) (not to be confused with Manning's n for open channel flow)

L = flow length, ft

P_2 = 2-yr 24-hr rainfall, in.

S = slope of flow path, ft/ft

If surface flow continues beyond sheet flow, it becomes *shallow concentrated flow*. Velocities for this type of flow

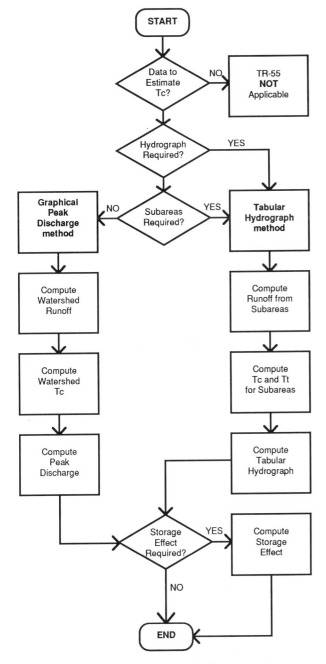

FIGURE 9.8. Selection Procedure Flow Chart.

TABLE 9.1. Roughness Coefficients (n) for Sheet Flow

Surface description	n[1]
Smooth surfaces (concrete, asphalt, gravel, or bare soil)	0.011
Fallow (no residue)	0.05
Cultivated soils:	
Residue cover ≤20%	0.06
Residue cover >20%	0.17
Grass:	
Short grass prairie	0.15
Dense grasses[2]	0.24
Bermudagrass	0.41
Range (natural)	0.13
Woods:[3]	
Light underbrush	0.40
Dense underbrush	0.80

[1]The n values are a composite of information compiled by Engman (1986).
[2]Includes species such as weeping lovegrass, bluegrass, buffalo grass, blue grama grass, and native grass mixtures.
[3]When selecting n, consider cover to a height of about 0.1 ft. This is the only part of the plant cover that will obstruct sheet flow.

runoff will probably be overland flow. The actual flow in the pipes must be determined separately.

Travel Time and Time of Concentration

Travel time (T_t) is the time for runoff to flow from one point in a drainage area to another; *time of concentration* (T_c) is the sum of the consecutive travel times through a watershed. (Previously this was defined as the time for water to flow from the hydraulically most remote point in a drainage area to the point of interest, usually the outlet.) With the flow distance and velocity of flow known, travel time can be calculated by dividing the flow length by the velocity, or

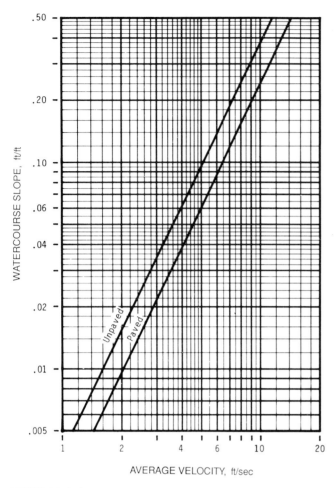

FIGURE 9.9. Average Velocities for Shallow Concentrated Flow.

TABLE 9.2. Roughness Coefficients (n) for Pipes and Channels

	n
Pipe Material	
Concrete	0.012–0.015
Cast Iron	0.013
Corrugated Metal	0.024
Plastic (smooth)	0.012
Vitrified Clay	0.010–0.015
Channel Surface	
Asphalt	0.013–0.016
Concrete	0.012–0.018
Riprap	0.020–0.040
Vegetated	0.030–0.080

200 ft of sheet flow through woods with light underbrush at a 4% slope

770 ft of shallow concentrated flow through woods at a 4% slope

1100 ft of open channel flow at a 2% slope

The cross-section of flow of the channel approximates a trapezoid with a top width of 16 ft, bottom width of 8 ft, and depth of 2 ft (Fig 9.11). The roughness coefficient has been estimated as 0.10. The 2-yr 24-hr rainfall in the area of the project is 3.3 in.

Solution. Sheet flow:

$$T_{t1} = \frac{0.007\,(nL)^{0.8}}{(P_2)^{0.5}S^{0.4}} = \frac{0.007\,(0.4 \times 200)^{0.8}}{3.3^{0.5}0.04^{0.4}} \approx 0.47 \text{ hr}$$

Shallow concentrated flow:

$V = 3.2$ fps (from Fig. 9.9)

$T_{t2} = 770 \text{ ft}/3.2 \text{ fps} = 240.6$ sec

$\phantom{T_{t2}} = 240.6 \text{ sec}/3600 \text{ sec/hr} \approx 0.07$ hr

Channel flow:

$$R = \frac{A}{WP}$$

$$A = \frac{B + b}{2} \times h = \frac{16 + 8}{2} \times 2 = 24 \text{ ft}^2$$

$WP = 2 \times$ length of sloped side $+$ base (b)

$ = 2 \times (4^2 + 2^2)^{0.5} + 8 = 2 \times 4.47 + 8 = 16.94$ ft

$$R = \frac{24 \text{ ft}^2}{16.94 \text{ ft}} \approx 1.42 \text{ ft}$$

$$V = \frac{1.486}{n}R^{2/3}S^{1/2} = \frac{1.486}{0.10} \times 1.42^{2/3} \times 0.02^{1/2} \approx 2.65 \text{ fps}$$

$$T_{t3} = \frac{1100 \text{ ft}}{2.65 \text{ fps}} \approx 415.1 \text{ sec}$$

$$\phantom{T_{t3}} = \frac{415.1 \text{ sec}}{3600 \text{ sec/hr}} \approx 0.12 \text{ hr}$$

$T_c = T_{t1} + T_{t2} + T_{t3} = 0.47 + 0.07 + 0.12 = 0.66$ hr

may be estimated by the use of Fig. 9.9 for paved or unpaved flow paths by entering with the slope in feet/foot (ft/ft) from the left until the appropriate line for type of flow is intersected. Projecting a vertical line down from the point of intersection, the velocity in feet per second (fps) may be read. Travel time can then be computed by means of Equation 9.1.

Finally, runoff may enter an *open channel*. The average flow velocity for open channels can be calculated by Manning's equation for open channels:

$$V = \frac{1.486}{n}R^{2/3}S^{1/2} \tag{9.3}$$

where: V = flow velocity, fps
             n = Manning's roughness coefficient for open channels (Table 9.2)
             R = hydraulic radius, ft (see Fig. 9.10)

$$ = \frac{\text{cross section area of flow, ft}^2}{\text{wetted perimeter, ft}}$$

             S = channel slope, ft/ft

Example 9.1

Calculate the time of concentration for the following conditions:

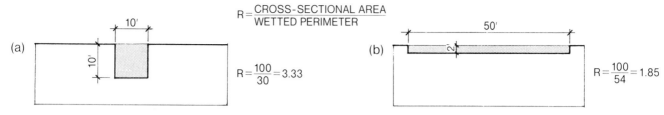

FIGURE 9.10. Hydraulic Radius. For the same given cross-sectional area, R varies inversely with the wetted perimeter. Although the cross-sectional area in (b) is the same as in (a), the surface area exposed per unit length of channel is much greater, resulting in greater friction between the channel and the water. Consistent with the increased friction, the R value is less than in (a) and the velocity of flow is reduced.

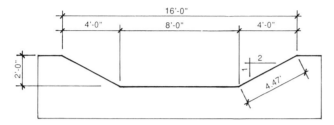

FIGURE 9.11. Cross Section of Flow for Example 9.1.

Computing Runoff

The basic runoff equation used by the SCS is:

$$Q = \frac{(P - I_a)^2}{(P - I_a) + S} \quad (9.4)$$

where: Q = runoff, in.
P = rainfall, in.
S = potential maximum retention after start of runoff, in.
I_a = initial abstraction, in.

The initial abstraction, that is, all losses before runoff begins, includes infiltration, evaporation, interception by vegetation, and water retained in surface depressions. Generally, initial abstraction is estimated as 0.2S. When this is substituted into the runoff equation, it becomes:

$$Q = \frac{(P - 0.2S)^2}{P + 0.8S} \quad (9.5)$$

S is defined as $1000/CN - 10$, where CN is a runoff curve number, which theoretically varies from 0 to 100 and is somewhat analogous to the runoff coefficient C in the Rational formula. Tables 9.3 and 9.4 list runoff curve numbers for different urban and natural land uses. TR55 has additional listings for various farming practices and range lands. When using these numbers, the proportion of impervious areas for the urban land uses and lot sizes on a project site should agree with those listed. Also, the numbers have been developed with the assumption that urban impervious areas are directly connected to a drainage system. If these assumptions do not apply, TR55 includes methods of adjusting the curve numbers for other conditions.

Hydrologic Soil Groups

The *hydrologic soil group (HSG)* classification indicates how much of the precipitation is likely to enter the soil (infiltrate) and how much will run off. HSG A includes sandy soils with high infiltration rates and relatively little runoff; HSG D soils, such as clay loams and silty clays, have very low infiltration and high runoff potential. HSG B and HSG C soils have intermediate characteristics.

Soil survey maps indicate the various soils existing on a project site. HSG classifications for all soils are listed in Appendix A of TR55 or may be obtained directly from soil survey reports. Normally, the project boundaries and drainage areas are drawn on the applicable soil map to the appropriate scale and the areas of the various soil types, with their HSG classification, contained within the drainage areas are measured. As with runoff coefficients for the Rational and Modified Rational methods, a weighted average CN may be calculated, where various soil types are intermingled on a site (Fig 9.12).

Hydrologic Condition

The term *hydrologic condition* applies to the vegetative cover, residue, and surface roughness of a particular area. Good plant cover and soil surfaces that are not smooth generally reduce potential runoff and cause a soil to be considered in good hydrologic condition.

Graphical Peak Discharge Method

The equation used for the graphical peak discharge method is:

$$q_p = q_u \times A_m \times Q \times F_p \quad (9.6)$$

where: q_p = peak discharge, cfs
q_u = unit peak discharge, cfs/square mile/in. of runoff (csm/in.)
A_m = drainage area, sq mi
Q = runoff, in.
F_p = pond and swamp adjustment factor (see Table 9.5) (do not include ponds or swamps that are in the T_c flow path)

This method can be used for homogeneous drainage areas that can be characterized by a single CN, if only peak discharge values are needed.

FIGURE 9.12. Plotting Hydrologic Soil Groups for a Project Site.

a. Soils map with project boundary indicated.

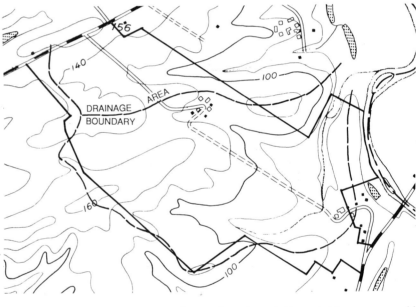

b. Topographic map with drainage area boundaries indicated.

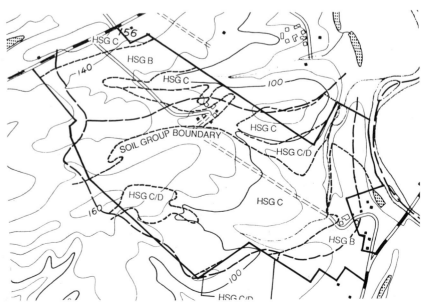

c. Boundaries of hydrologic soil groups (HSG) superimposed on topographic map with drainage areas.

TABLE 9.3. Runoff Curve Numbers for Urban Areas[1]

Cover description		Curve numbers for hydrologic soil group—			
Cover type and hydrologic condition	Average percent impervious area[2]	A	B	C	D

Fully developed urban areas (vegetation established)

Open space (lawns, parks, golf courses, cemeteries, etc.)[3]:

Poor condition (grass cover < 50%)		68	79	86	89
Fair condition (grass cover 50% to 75%)		49	69	79	84
Good condition (grass cover > 75%)		39	61	74	80

Impervious areas:

Paved parking lots, roofs, driveways, etc. (excluding right-of-way)		98	98	98	98

Streets and roads:

Paved; curbs and storm sewers (excluding right-of-way)		98	98	98	98
Paved; open ditches (including right-of-way)		83	89	92	93
Gravel (including right-of-way)		76	85	89	91
Dirt (including right-of-way)		72	82	87	89

Western desert urban areas:

Natural desert landscaping (pervious areas only)[4]		63	77	85	88
Artificial desert landscaping (impervious weed barrier, desert shrub with 1- to 2-inch sand or gravel mulch and basin borders)		96	96	96	96

Urban districts:

Commercial and business	85	89	92	94	95
Industrial	72	81	88	91	93

Residential districts by average lot size:

1/8 acre or less (town houses)	65	77	85	90	92
1/4 acre	38	61	75	83	87
1/3 acre	30	57	72	81	86
1/2 acre	25	54	70	80	85
1 acre	20	51	68	79	84
2 acres	12	46	65	77	82

Developing urban areas

Newly graded areas (pervious areas only, no vegetation)[5]		77	86	91	94
Idle lands (CN's are determined using cover types similar to those in table 2-2c).					

[1] Average runoff condition, and $I_a = 0.2S$.
[2] The average percent impervious area shown was used to develop the composite CN's. Other assumptions are as follows: impervious areas are directly connected to the drainage system, impervious areas have a CN of 98, and pervious areas are considered equivalent to open space in good hydrologic condition. CN's for other combinations of conditions may be computed using figure 2-3 or 2-4.
[3] CN's shown are equivalent to those of pasture. Composite CN's may be computed for other combinations of open space cover type.
[4] Composite CN's for natural desert landscaping should be computed using figures 2-3 or 2-4 based on the impervious area percentage (CN = 98) and the pervious area CN. The pervious area CN's are assumed equivalent to desert shrub in poor hydrologic condition.
[5] Composite CN's to use for the design of temporary measures during grading and construction should be computed using figure 2-3 or 2-4, based on the degree of development (impervious area percentage) and the CN's for the newly graded pervious areas.

TABLE 9.4. Runoff Curve Numbers for Other Agricultural Lands[1]

Cover description		Curve numbers for hydrologic soil group—			
Cover type	Hydrologic condition	A	B	C	D
Pasture, grassland, or range—continuous forage for grazing.[2]	Poor	68	79	86	89
	Fair	49	69	79	84
	Good	39	61	74	80
Meadow—continuous grass, protected from grazing and generally mowed for hay.	—	30	58	71	78
Brush—brush-weed-grass mixture with brush the major element.[3]	Poor	48	67	77	83
	Fair	35	56	70	77
	Good	*30	48	65	73
Woods—grass combination (orchard or tree farm).[5]	Poor	57	73	82	86
	Fair	43	65	76	82
	Good	32	58	72	79
Woods.[6]	Poor	45	66	77	83
	Fair	36	60	73	79
	Good	*30	55	70	77
Farmsteads—buildings, lanes, driveways, and surrounding lots.	—	59	74	82	86

[1] Average runoff condition, and I_a = 0.2S.

[2] *Poor:* <50% ground cover or heavily grazed with no mulch.
Fair: 50 to 75% ground cover and not heavily grazed.
Good: >75% ground cover and lightly or only occasionally grazed.

[3] *Poor:* <50% ground cover.
Fair: 50 to 75% ground cover.
Good: >75% ground cover.

[4] Actual curve number is less than 30; use CN = 30 for runoff computations.

[5] CN's shown were computed for areas with 50% woods and 50% grass (pasture) cover. Other combinations of conditions may be computed from the CN's for woods and pasture.

[6] *Poor:* Forest litter, small trees, and brush are destroyed by heavy grazing or regular burning.
Fair: Woods are grazed but not burned, and some forest litter covers the soil.
Good: Woods are protected from grazing, and litter and brush adequately cover the soil.

TABLE 9.5. Pond and Swamp Adjustment Factor (F_p)

Percentage of pond and swamp areas	F_p
0	1.00
0.2	0.97
1.0	0.87
3.0	0.75
5.0	0.72

Example 9.2

Determine the 50-yr 24-hr peak flow rate for a 95-ac wooded drainage area in central New Jersey. The soil on the site is classified as Bucks silt loam in fair hydrologic condition. The time of concentration has been computed as 0.66 hr (see Example 9.1). There are no ponds or swamps.

Solution. Appendix A in TR55 lists Bucks silt loam as HSG B. The CN for HSG B woods in fair condition is 60 from Table 9.4.

A_m = 95 ac/640 ac/sq mi = 0.148 sq mi

P = 6.5 in (50-yr 24-hr precipitation in central New Jersey from Fig. 9.6)

S = (1000/CN) − 10 = (1000/60) − 10 = 6.67 in.

I_a = 0.2S = 0.2 × 6.67 in. = 1.33 in.

$Q = \dfrac{(P - 0.2S)^2}{P + 0.8S} = \dfrac{(6.5 - 1.33)^2}{6.5 + (0.8 \times 6.67)} = 2.26$ in.

I_a/P = 1.33/6.5 = 0.20

Using Fig. 9.13 for type III rainfall, 0.66-hr T_c, and I_a/P = 0.20, q_u is estimated as 330 csm. Finally:

$$q_p = q_u \times A_m \times Q \times F_p$$
$$= 330 \times 0.148 \times 2.26 \times 1.0 = 110.38 \approx 110 \text{ cfs}$$

Note that approximate Q values may also be obtained directly from Fig. 9.1.

Tabular Hydrograph Method

When a hydrograph is required or when a drainage area consists of several distinct subareas with different land uses, topography, or soils, the use of the Tabular Hydrograph method is required as shown in Fig. 9.8. The acreage of the various subareas should not differ by a factor of 5 or more. The equation used for the flow rate at any time during runoff is:

$$q = q_t \times A_m \times Q \qquad (9.7)$$

where: q = hydrograph coordinate at hydrograph time t, cfs
q_t = tabular hydrograph unit discharge from the tables provided in TR55, csm/in. of runoff
A_m = subarea, sq mi
Q = runoff, in.

Example 9.3

Develop a combined 10-yr, 24-hr hydrograph for the grassed and paved areas of Example 8.6. Properties of the areas are as follows:

Area A: Lawn, fair condition, HSG B

$CN = 69$ (from Table 9.3)
$T_c = 0.5$ hr
$T_t = 0.1$ hr (over pavement)

Area B: Paved

$CN = 98$ (for all HSGs) (from Table 9.3)
$T_c = 0.1$ hr
$T_t = 0.0$ hr

The 10-yr, 24-hr precipitation (central New Jersey) is 5.2 in. from Fig. 9.5. A type III storm distribution is to be used.

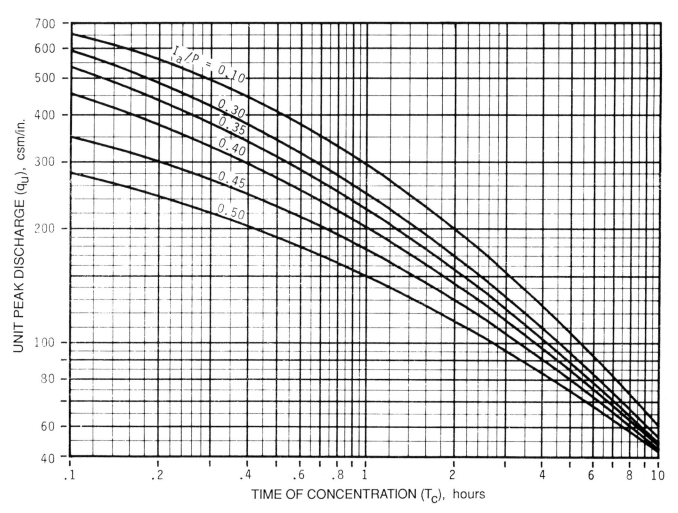

FIGURE 9.13. Unit Peak Discharge (q_u) for SCS Type III Rainfall Distribution.

Solution.

Area *A*

$S = (1000/69) - 10 = 4.49$ in.

$I_a = 0.2 \times 4.49 = 0.898 \approx 0.9$ in.

$I_a/P = 0.9/5.2 = 0.17$

$Q = \dfrac{(5.2 - 0.9)^2}{5.2 + (0.8 \times 4.49)} = 2.10$ in.

$A_m = 3/640 = 0.0047$ sq mi

Area *B*

$S = (1000/98) - 10 = 0.20$ in.

$I_a = 0.2 \times 0.2 = 0.04$ in.

$I_a/P = 0.04/5.2 = 0.0077$

$Q = \dfrac{(5.2 - 0.04)^2}{5.2 + (0.8 \times 0.2)} = 4.967 \approx 5.0$ in.

$A_m = 0.0047$ sq mi

Hydrograph for Area *A*

Area *A* has a T_c of 0.5 hr and a T_t of 0.1 hr. The I_a/P values for which hydrographs are listed in TR55 are 0.10, 0.30, and 0.50. Either the computed I_a/P value can be rounded to the nearest one of these three values or the runoff ordinates can be linearly interpolated. For this example the computed I_a/P of 0.17 will be rounded down to 0.10.

Line 1 in Table 9.6 lists the hydrograph time; line 2 the unit discharge, q_t, in csm/in. at that time; and line 3 the discharge, q, at that time in cfs. Lines 1 and 2 have been obtained from Table 9.9b for $T_c = 0.5$ hr and $T_t = 0.1$ hr. Line 3 was computed by multiplying the csm/in. values by A_m, the area in square miles, and by Q, the runoff in inches. A_m had previously been determined as 0.0047 and Q as 2.10. These values can be combined as a multiplication factor of 0.0099. All cfs values have been rounded to the nearest 0.1 cfs.

Hydrograph for Area *B*

Similarly, using Table 9.9a with $T_c = 0.1$ hr, $T_t = 0.0$ hr, $I_a/P = 0.10$, $A_m = 0.0047$ sq mi, and $Q = 5.0$ in. (multiplication factor of 0.0235), cfs values are computed as shown in Table 9.7.

The two hydrographs can now be combined by adding, as shown in Table 9.8. The combined 10-yr 24-hr hydrograph yields q_p of 16.5 cfs at 12.2 hr (Table 9.8). The timing of the peak flow reflects the 24-hr rainfall distribution for a type III storm and a short time of concentration.

VOLUME FOR DETENTION STORAGE

Figure 9.14 provides a method of estimating required detention storage volume when the postdevelopment peak runoff rate from a site into a basin and the limiting peak outflow rate from the basin are known. As with the previously described Modified Rational method, the detention volume obtained by this procedure should normally not be used for a final design but can aid in the selection of the pond location and provide a starting point for a detailed design. An accurate stage-volume relationship for a basin and the details of the outlet structure must be known to perform the reservoir routing for the final design. A discussion of that procedure is not within the scope of this text.

To use Fig. 9.14 the TR55 graphical peak discharge method or the tabular hydrograph method *must* be used. If the pond is to be located at the outlet of a subarea, travel times downstream of the subarea should not be included in the determination of the peak flow rate from the subarea.

Example 9.4

The 95-ac wooded area of Example 9.2 is to be developed into 1-ac residential lots. Determine the approximate pond storage required to maintain the 110 cfs predevelopment outflow rate for a 50-yr 24-hr storm event (previously computed). The postdevelopment 50-yr 24-hr peak runoff rate has been determined as 183 cfs.

Solution. The ratio of peak outflow rate/peak inflow rate equals 110 cfs/183 cfs = 0.60. Drawing a vertical line upward on Fig. 9.13 from 0.60 on the *x* axis to the type III curve (New Jersey) and then a horizontal line to the left, a storage volume/runoff volume (V_s/V_r) ratio of about 0.24 is obtained.

$P = 6.5$ in. for a 50-yr 24-hr storm event (central New Jersey)

$CN = 68$ for 1-ac lots on HSG B soil (Table 9.3)

$Q = 3.1$ from Fig. 9.1

The volume of runoff (V_r) is computed as follows:

$V_r = 3.1$ in. $\times 95$ ac $= 294.5$ ac-in.

$ = 294.5$ ac-in./12 in./ft $= 24.54$ ac-ft

The required storage volume (V_s) can then be determined by using the V_s/V_r ratio obtained from Fig. 9.13.

$\dfrac{V_s}{V_r} = 0.24$

$V_s = V_r \times 0.24 = 24.54$ ac-ft $\times 0.24 = 5.89 \approx 5.9$ ac-ft

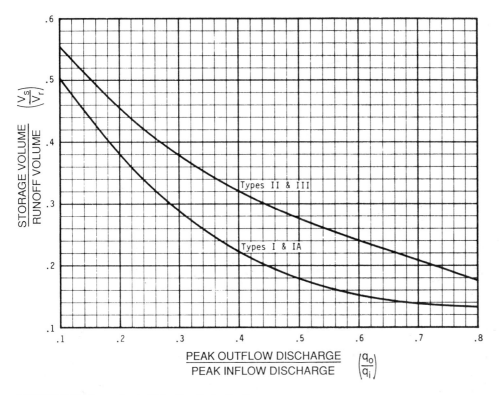

FIGURE 9.14. Approximate Detention Basin Routing.

TABLE 9.6. Area A Hydrograph for Example 9.3

hr[a]	11.0	11.3	11.6	11.9	12.0	12.1	12.2	12.3	12.4	12.5	12.6	12.7	12.8	13.0	13.2	13.4	13.6	13.8	14.0
csm/in.[b]	19.0	24.0	30.0	43.0	50.0	64.0	86.0	125.0	186.0	273.0	355.0	392.0	390.0	296.0	194.0	129.0	94.0	75.0	65.0
cfs[c]	0.2	0.2	0.3	0.4	0.5	0.6	0.9	1.2	1.8	2.7	3.5	3.9	3.9	2.9	1.9	1.3	0.9	0.7	0.6
hr	14.3	14.6	15.0	15.5	16.0	16.5	17.0	17.5	18.0	19.0	20.0	22.0	26.0						
csm/in.	56.0	49.0	43.0	38.0	33.0	28.0	24.0	21.0	19.0	15.0	14.0	11.0	0.0						
cfs	0.6	0.5	0.4	0.4	0.3	0.3	0.2	0.2	0.2	0.1	0.1	0.1	0.0						

[a] Hydrograph time
[b] Unit discharge, q_t, from Table 9.9b for $t_c = 0.5$ hr, $T_t = 0.1$ hr, and $I_a/P = 0.10$
[c] From $q = q_t \times A_m \times Q$

TABLE 9.7. Area B Hydrograph for Example 9.3

hr[a]	11.0	11.3	11.6	11.9	12.0	12.1	12.2	12.3	12.4	12.5	12.6	12.7	12.8	13.0	13.2	13.4	13.6	13.8	14.0
csm/in.[b]	29.0	38.0	57.0	172.0	241.0	425.0	662.0	531.0	345.0	265.0	191.0	130.0	101.0	83.0	68.0	62.0	58.0	54.0	50.0
cfs[c]	0.7	0.9	1.3	4.0	5.7	10.0	15.6	12.5	8.1	6.2	4.5	3.1	2.4	2.0	1.6	1.5	1.4	1.3	1.2
hr.	14.3	14.6	15.0	15.5	16.0	16.5	17.0	17.5	18.0	19.0	20.0	22.0	26.0						
csm/in.	44.0	41.0	37.0	32.0	27.0	23.0	21.0	19.0	16.0	14.0	13.0	11.0	0.0						
cfs	1.0	1.0	0.9	0.8	0.6	0.5	0.5	0.4	0.4	0.3	0.3	0.3	0.0						

[a] Hydrograph time
[b] Unit discharge, q_t, from Table 9.9a for $t_c = 0.1$ hr, $T_t = 0.0$ hr, and $I_a/P = 0.10$
[c] From $q = q_t \times A_m \times Q$

TABLE 9.8. Combined Hydrograph for Example 9.3

hr	11.0	11.3	11.6	11.9	12.0	12.1	12.2	12.3	12.4	12.5	12.6	12.7	12.8	13.0	13.2	13.4	13.6	13.8	14.0
A cfs	0.2	0.2	0.3	0.4	0.5	0.6	0.9	1.2	1.8	2.7	3.5	3.9	3.9	2.9	1.9	1.3	0.9	0.7	0.6
B cfs	0.7	0.9	1.3	4.0	5.7	10.0	15.6	12.5	8.1	6.2	4.5	3.1	2.4	2.0	1.6	1.5	1.4	1.3	1.2
Total cfs	0.9	1.1	1.6	4.4	6.2	10.6	16.5	13.7	9.9	8.9	8.0	7.0	6.3	4.9	3.5	2.8	2.3	2.0	1.8
hr	14.3	14.6	15.0	15.5	16.0	16.5	17.0	17.5	18.0	19.0	20.0	22.0	26.0						
A cfs	0.6	0.5	0.4	0.4	0.3	0.3	0.2	0.2	0.2	0.1	0.1	0.1	0.0						
B cfs	1.0	1.0	0.9	0.8	0.6	0.5	0.5	0.4	0.4	0.3	0.3	0.3	0.0						
Total cfs	1.6	1.5	1.3	1.2	0.9	0.8	0.7	0.6	0.6	0.4	0.4	0.4	0.0						

TABLE 9.9a. Tabular Hydrograph Unit Discharges (csm/in.) for Type III Rainfall Distribution ($T_c = 0.1$ hr)

TRVL TIME (HR)	11.0	11.3	11.6	11.9	12.0	12.1	12.2	12.3	12.4	12.5	12.6	12.7	12.8	13.0	13.2	13.4	13.6	13.8	14.0	14.3	14.6	15.0	15.5	16.0	16.5	17.0	17.5	18.0	19.0	20.0	22.0	26.0
																	TC =0.1 HR												**IA/P = 0.10**				
0.0	29	38	57	172	241	425	662	531	345	265	191	130	101	83	68	62	58	54	50	44	41	37	32	27	23	21	19	16	14	13	11	0
.10	26	32	47	98	147	210	353	559	540	410	313	231	164	101	80	67	61	57	53	47	43	39	34	29	24	22	19	17	14	13	11	0
.20	25	31	44	86	127	182	296	471	517	446	357	273	200	117	86	70	63	58	54	48	44	39	34	29	24	22	20	17	15	13	11	0
.30	22	28	37	57	76	110	158	250	398	477	457	390	312	178	111	83	69	62	57	51	45	41	36	31	25	23	20	18	15	13	11	0
.40	21	27	35	53	68	96	137	213	336	430	448	410	345	210	128	90	72	64	58	52	46	41	36	31	26	23	20	18	15	13	11	2
.50	19	24	30	43	62	85	120	182	284	382	426	415	375	305	188	120	86	71	63	55	49	43	38	33	27	24	21	19	16	14	12	6
.75	17	22	27	37	49	62	84	120	181	258	327	375	353	264	177	120	88	72	63	52	48	45	39	34	29	25	22	20	19	16	13	9
1.0	13	17	22	30	33	37	43	52	66	91	131	190	220	358	307	220	149	104	57	60	52	50	43	37	32	27	23	21	17	15	14	10
1.5	9	11	14	18	19	21	23	25	27	29	33	37	44	70	134	229	304	318	269	172	106	68	52	44	38	33	28	24	19	15	12	2
2.0	6	8	10	13	14	15	16	17	19	20	24	26	32	45	73	130	207	271	271	292	216	121	68	51	43	37	32	27	21	16	13	6
2.5	3	4	6	8	10	11	12	13	14	13	16	17	20	23	29	38	57	97	189	271	244	136	75	53	44	32	38	33	27	21	14	9
3.0	1	2	4	5	6	7	8	8	9	9	10	11	12	14	16	19	23	28	38	74	146	256	226	131	74	53	44	37	27	21	14	10
																													IA/P = 0.30				
0.0	0	0	0	0	48	106	296	496	368	300	221	155	125	106	89	83	79	74	69	62	59	54	47	40	35	32	28	25	22	20	17	0
.10	0	0	0	0	35	82	225	488	408	336	260	190	147	113	94	85	80	75	70	63	59	54	48	40	35	32	28	25	22	20	17	0
.20	0	0	0	0	26	64	171	372	449	422	365	295	225	142	102	92	84	79	74	66	61	56	50	43	36	33	30	26	22	20	17	0
.30	0	0	0	0	5	19	49	130	291	397	414	381	323	161	118	96	86	80	75	68	62	57	50	43	37	33	30	27	22	20	17	0
.40	0	0	0	0	3	14	37	99	227	340	389	384	343	229	152	113	94	85	79	71	65	59	52	46	38	34	31	28	23	21	17	0
.50	0	0	0	0	2	10	28	75	177	286	355	374	354	256	170	123	99	87	80	73	66	60	53	46	39	35	31	28	24	21	18	0
.75	0	0	0	0	0	0	4	13	35	86	238	295	325	266	194	141	110	93	80	71	63	56	50	43	37	33	30	26	22	20	18	2
1.0	0	0	0	0	0	0	0	2	6	19	48	99	194	282	264	197	144	112	97	88	77	67	59	52	45	39	34	31	27	22	18	2
1.5	0	0	0	0	0	0	0	0	0	0	1	0	4	29	99	197	265	277	236	162	113	84	69	60	53	46	39	35	28	23	19	2
2.0	0	0	0	0	0	0	0	0	0	0	0	1	0	1	8	35	94	172	233	253	196	124	83	68	59	52	45	39	31	25	20	8
2.5	0	0	0	0	0	0	0	0	0	0	0	0	0	0	2	11	37	88	184	235	201	122	67	59	52	45	39	34	27	21	15	13
3.0	0	0	0	0	0	0	0	0	0	0	0	0	0	0	0	0	1	7	22	110	202	222	131	52	60	52	45	39	31	22	17	15
																													IA/P = 0.50				
0.0	0	0	0	0	0	0	0	107	258	209	155	130	107	97	91	87	82	78	74	69	65	61	52	47	43	39	35	29	25	0
.10	0	0	0	0	0	0	0	71	282	224	178	146	112	99	92	88	83	80	75	70	66	62	53	48	44	40	35	32	29	25	0
.20	0	0	0	0	0	0	0	48	246	239	208	162	127	109	95	91	87	83	77	72	65	56	49	45	41	36	32	30	26	0
.30	0	0	0	0	0	0	0	32	132	216	225	195	162	113	101	96	91	88	81	77	72	65	57	50	45	41	36	32	30	26	0
.40	0	0	0	0	0	0	0	0	21	73	139	213	208	164	131	111	100	95	91	85	79	74	68	60	51	47	43	38	33	30	26	1
.50	0	0	0	0	0	0	0	0	14	53	110	164	213	174	139	116	103	97	92	86	80	75	68	60	52	48	44	39	33	30	26	1
.75	0	0	0	0	0	0	0	0	5	20	54	96	137	180	159	134	115	103	96	89	83	77	70	63	54	48	44	40	33	31	27	7
1.0	0	0	0	0	0	0	0	0	0	2	10	29	60	132	169	146	124	109	97	81	74	67	59	52	47	43	34	31	27	7
1.5	0	0	0	0	0	0	0	0	0	0	0	0	0	2	17	58	112	159	148	122	104	91	81	74	67	59	51	46	38	32	28	1
2.0	0	0	0	0	0	0	0	0	0	0	0	0	0	0	0	20	54	98	133	149	133	108	90	80	73	66	58	51	42	34	29	7
2.5	0	0	0	0	0	0	0	0	0	0	0	0	0	0	0	4	12	35	87	141	161	141	111	92	81	73	66	59	46	38	30	18
3.0	0	0	0	0	0	0	0	0	0	0	0	0	0	0	0	0	4	22	63	120	136	110	91	81	73	66	51	42	31	22

RAINFALL TYPE = III

TABLE 9.9b. Tabular Hydrograph Unit Discharges (csm/in.) for Type III Rainfall Distribution ($T_c = 0.5$ hr)

TRVL TIME (HR)	11.0	11.3	11.6	11.9	12.0	12.1	12.2	12.3	12.4	12.5	12.6	12.7	12.8	13.0	13.2	13.4	13.6	13.8	14.0	14.3	14.6	15.0	15.5	16.0	16.5	17.0	17.5	18.0	19.0	20.0	22.0	26.0

*** TC = 0.5 HR *** IA/P = 0.10

0.0	21	27	35	54	70	97	144	217	316	397	388	330	214	139	99	78	67	60	52	47	42	36	31	26	23	21	18	15	13	11	0
.10	19	24	30	43	50	64	86	125	186	273	355	390	296	194	129	94	75	65	56	49	43	38	33	28	24	21	19	15	14	11	0
.20	18	23	29	40	47	58	77	109	161	235	315	367	318	218	145	103	80	68	57	50	44	39	33	28	24	21	19	15	14	11	0
.30	16	21	26	34	38	44	53	69	95	139	203	278	337	289	199	135	98	77	62	54	46	40	35	30	25	22	20	16	14	11	0
.40	16	20	25	33	36	41	49	62	84	121	176	244	306	358	220	151	107	83	64	55	47	41	35	30	25	23	20	16	14	12	0
.50	14	18	22	28	31	35	39	46	57	75	106	152	213	323	282	202	140	102	73	59	50	42	37	32	27	21	18	12	0
.75	12	16	20	25	28	30	34	38	45	56	75	104	145	246	308	252	187	135	89	67	53	44	39	33	28	25	22	17	14	12	0
1.0	10	12	16	20	22	23	25	.28	31	34	39	47	60	110	197	279	220	138	98	63	49	42	37	31	26	23	20	16	15	12	1
1.5	6	8	10	13	14	15	17	18	19	21	23	25	27	34	49	82	143	218	283	271	203	116	68	51	43	37	32	27	21	16	13	4
2.0	3	5	7	9	10	11	12	13	14	15	16	17	19	22	27	34	50	82	135	202	226	211	114	67	50	42	37	31	23	18	13	8
2.5	2	3	4	6	7	7	8	9	10	11	12	13	16	18	22	26	34	50	102	182	249	197	111	67	50	42	36	26	20	14	9
3.0	1	2	3	4	4	4	5	5	6	6	7	8	10	12	14	16	19	23	34	63	144	238	201	121	72	52	43	31	23	15	10

IA/P = 0.30

0.0	0	0	0	0	0	0	0	0	0	0	40	101	198	295	345	325	232	161	122	100	88	80	72	65	59	53	46	39	34	31	28	23	21	18	0
.10	0	0	0	0	0	0	3	11	30	77	158	249	313	335	253	178	132	106	91	82	73	66	60	54	47	40	35	31	28	23	21	18	0
.20	0	0	0	0	0	0	2	8	23	59	125	208	278	316	271	196	144	112	95	85	75	67	61	54	47	40	35	28	23	21	18	0
.30	0	0	0	0	0	0	0	0	6	17	45	98	171	242	291	313	249	182	136	108	92	80	71	63	56	49	42	36	29	24	21	18	0
.40	0	0	0	0	0	0	1	4	13	34	77	140	208	263	304	295	244	185	148	115	97	81	72	64	57	50	43	37	30	24	21	15	0
.50	0	0	0	0	0	0	1	3	10	26	60	113	178	263	291	263	174	140	111	88	77	67	59	52	45	39	34	27	22	19	15	0
.75	0	0	0	0	0	0	0	1	4	12	29	60	104	204	271	263	222	174	136	101	83	70	61	54	47	40	35	32	25	22	18	0
1.0	0	0	0	0	0	0	0	0	1	2	6	16	29	67	155	242	263	198	138	102	92	80	71	63	58	51	44	38	27	23	19	1
1.5	0	0	0	0	0	0	0	0	0	0	1	4	16	67	155	205	241	221	167	110	79	66	58	51	44	38	30	24	21	18	5
2.0	0	0	0	0	0	0	0	0	0	0	0	0	4	22	67	138	205	241	182	119	83	67	58	51	44	34	27	22	18	5
2.5	0	0	0	0	0	0	0	0	0	0	0	0	0	3	13	42	93	182	225	139	95	70	61	54	43	36	28	22	15
3.0	0	0	0	0	0	0	0	0	0	0	0	0	0	0	0	4	15	62	139	127	203	182	130	105	80	61	47	33	29	15	16

*** TC = 0.5 HR *** IA/P = 0.50

0.0	0	0	0	0	0	0	3	24	68	124	174	190	162	133	114	103	97	92	85	80	75	68	60	52	47	43	39	33	30	26	0
.10	0	0	0	0	0	0	2	17	51	100	149	174	169	140	119	106	99	93	86	81	75	69	61	52	47	43	39	33	30	26	0
.20	0	0	0	0	0	0	1	12	38	79	126	160	173	147	124	109	101	95	88	81	76	69	62	53	48	44	39	33	30	26	0
.30	0	0	0	0	0	0	0	1	8	28	62	105	141	176	165	147	120	107	99	91	84	78	71	64	56	49	45	41	31	26	0
.40	0	0	0	0	0	0	0	6	20	48	86	123	172	157	125	111	101	92	85	79	72	65	56	50	45	41	34	31	26	0
.50	0	0	0	0	0	0	0	4	15	37	70	105	172	157	130	114	104	94	87	79	73	66	57	50	46	42	34	31	27	0
.75	0	0	0	0	0	0	0	0	1	6	23	48	91	157	167	150	119	103	93	84	76	69	62	54	48	44	35	32	27	0
1.0	0	0	0	0	0	0	0	0	0	0	1	17	37	91	135	153	135	120	107	99	91	88	79	72	65	57	50	45	37	32	27	1
1.5	0	0	0	0	0	0	0	0	0	0	0	0	5	24	59	101	132	144	130	107	90	80	73	65	57	51	41	34	29	6
2.0	0	0	0	0	0	0	0	0	0	0	0	0	0	0	5	24	59	101	132	144	138	105	89	79	72	65	57	45	37	30	15
2.5	0	0	0	0	0	0	0	0	0	0	0	0	0	0	0	7	25	55	138	133	133	104	88	71	64	50	41	31	21
3.0	0	0	0	0	0	0	0	0	0	0	0	0	0	0	0	0	0	9	36	91	138	133	128	103	88	70	55	45	32	23

*** TC = 0.5 HR ***

RAINFALL TYPE = III

If, for planning purposes, an average depth of 4 ft is assumed for the storage facility, then an area of approximately 1.47 ac (5.9 ac-ft/4 ft) would be required for the pond.

SUMMARY

Soil Conservation Service Technical Release 55 methods for calculating time of concentration and travel time, computing peak runoff rates, developing hydrographs, and estimating required detention storage have been presented in this chapter. Although examples of the various procedures have been shown, numerous graphs and tables from the source document are not included in this text and the reader is urged to obtain the original publication if the procedures are to be applied to an actual project.

EXERCISES

9.1 A 60-ac site in central New Jersey, presently in light woods in fair hydrologic condition, is to be developed into ½-ac residential lots. A soil survey indicates that 60% of the soils on the site are classified as HSG C and 40% as HSG B. There are no swamps or ponds on the site. The 2-yr and 100-yr 24-hr rainfall amounts are 3.3 in. and 7.5 in. respectively. The predevelopment flow path for determining the T_c is as follows:

290 ft of sheet flow with a 3% slope
500 ft of shallow concentrated flow with a 3.5% slope
1800 ft of weedy ditch at a 1% slope

The ditch has an n value of 0.13, a 10-ft bottom width, a 3.4-ft flow depth, and 2:1 *(H:V)* side slopes. The postdevelopment T_c will be one-half that of the predevelopment time.

a. Determine the predevelopment and postdevelopment times of concentration.
b. Determine the predevelopment 100-yr 24-hr peak rate of runoff.
c. Develop a hydrograph for the postdevelopment conditions.
d. Estimate the volume of detention storage needed for a pond to reduce the postdevelopment 100-yr 24-hr peak runoff rate to the corresponding predevelopment rate.

10
Design and Sizing of Storm Water Management Systems

MANAGEMENT SYSTEMS

The purpose of managing runoff is to ameliorate safety and health hazards, including flooding and property damage, stagnation, earth slides, and reduced soil-bearing capacity; to increase the usability of areas through the elimination of unwanted water; to provide better growing conditions for plants by increasing soil aeration and reducing soil saturation; and to prevent erosion by reducing the rate of flow and volume of runoff. There are a variety of management techniques that may be used to control storm water runoff. The context, that is, the purpose and environmental conditions, influence the selection of appropriate techniques.

The selection of an appropriate drainage system is based on a variety of factors. The scale and intensity of development, amount and location of paved and unpaved surfaces, proposed uses, ecological impacts, and aesthetic concerns must be addressed in making a choice. Physical factors, such as soil erodibility, extent and steepness of slopes, and expected rainfall intensities, must also be considered. In addition, the availability and suitability of a potential drainage outlet and the character of existing local systems may help determine the type of proposed system. Building and environmental codes or other regulatory requirements, such as maximum rate and volume of runoff, water quality, and method of connection to an existing system, must be met. The ultimate objective in designing most storm drainage systems is not to exceed the rate of flow that existed before the development of a site, for all storm frequencies.

Components of Storm Water Management Systems

Some components of storm water management systems have already been presented in Chapter 7. Other components common to management systems, as well as a review of those previously discussed, are presented here.

Swale
A constructed or natural drainage channel used to direct surface flow. Constructed swales have parabolic, trapezoidal, or triangular cross sections.

Culvert
Any structure not classified as a bridge, which allows a watercourse to flow beneath a road, walk, or highway. A *pipe culvert* has a circular, elliptical, or arched cross section; a *box culvert* has a rectangular cross section.

Catch Basin
A structure, typically concrete block or precast concrete rings, 2.5 to 4 ft in diameter, used to collect and divert surface runoff to an underground conduction system. A general rule of thumb is that one catch basin may be used for each 10,000 ft^2 of paved surface. At the base of the catch basin is a sump or sediment bowl to trap and collect debris. Catch basins may also be rectangular.

Drain Inlet
A structure that allows surface runoff to enter directly into a drain pipe. It does not contain a sump.

Area Drain
A prefabricated structure, similar to a floor drain, that collects runoff from paved areas. Usually one is used for each 1000 to 2000 ft^2 of pavement.

Trench Drain
A linear inlet structure used to collect sheet flow runoff in paved areas.

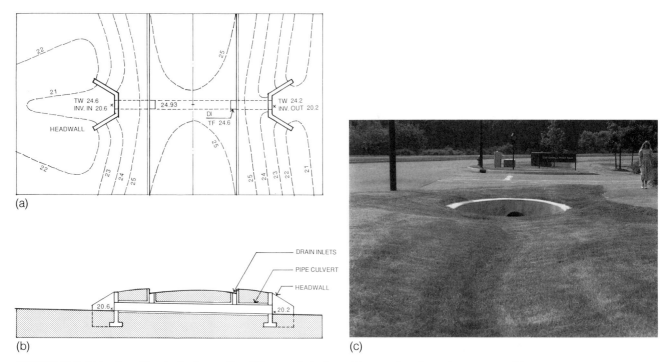

FIGURE 10.1. Culvert. a. Plan. b. Section. c. Runoff from swale is directed into a culvert beneath an entrance drive.

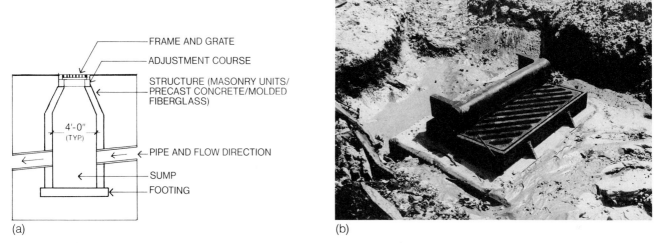

FIGURE 10.2. Catch Basin. a. Section. b. Catch basin with curb inlet under construction.

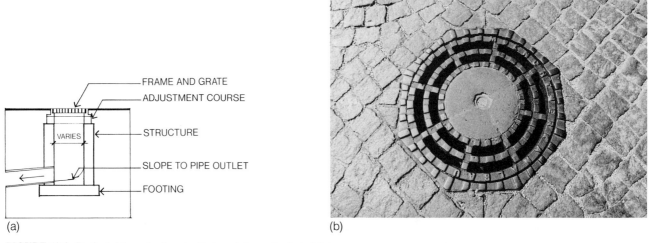

FIGURE 10.3. Drain Inlet. a. Section. b. Grate and frame for drain inlet.

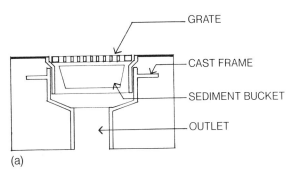

FIGURE 10.4. Area Drain. a. Section. b. Area drain grates. Notice how the placement of the grates is coordinated with the paving pattern.

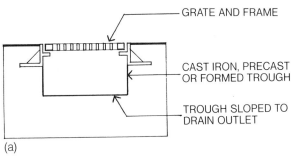

FIGURE 10.5. Trench Drain. a. Section. b. Trench drain under construction. A prepared base course (see Chapter 11) and grade stake are also illustrated.

Manhole
A structure, often 4 ft in diameter, made of concrete block, precast concrete, or fiberglass-reinforced plastic rings, which allows a person to enter a space below ground. For storm drainage purposes, manholes are used where there is a change in size, slope, or direction of underground pipes. Combining catch basins and manholes is cost-effective. Manholes may be rectangular or circular.

Detention and Retention Basin
An impoundment area constructed to collect storm runoff from a management system for the purpose of reducing peak flow and controlling rate of flow. A *retention basin* may be defined as having a permanent pool, whereas a *detention basin* is normally dry. (See Chapter 7.)

Sediment Basin
An impoundment area or structure that slows the velocity of runoff to allow sediment particles to settle out. Retention basins also function as sediment basins, although the reverse is not necessarily true. Retention, detention, and sediment basins require periodic cleaning to remove sediment.

Infiltration Basin
An open surface storage area with no outlet except an emergency spillway. (See Chapter 7.)

Open Drainage System

In an open drainage system (Fig. 10.6a) all surface runoff from paved and unpaved areas is collected and conveyed on the ground, primarily by swales. The system is discharged or directed to an on- or off-site drainageway, stream, or other natural watercourse; an existing street or municipal storm drainage system; or an on-site retention or sediment pond. The components of a system may include swales, gutters, channels, culverts, and detention, retention, sediment, or infiltration basins. In designing the system, consideration must be given to the volume and velocity of runoff to prevent swale erosion, and to the means of controlling discharge at the outlet in order to collect sediment and, if necessary, to dissipate flow energy to prevent erosion.

Closed Drainage System

In a closed drainage system, surface runoff from paved and unpaved areas is collected at surface inlets and conveyed by underground pipes to an outlet either on or off the site (Fig. 10.6b). The advantage of this system is that the runoff may be intercepted before the volume and velocity increase to the point of causing erosion. Disadvantages include increased cost and complexity of the system, potential erosion at the discharge point caused by the greater concentration of runoff, reduced filtering of sediment because of increased velocity of the storm water in the pipes, and decreased opportunity for storm water to infiltrate the soil. Structures commonly associated with closed systems are catch basins, drain inlets, area drains, trench drains, manholes, and piping. Piping materials include reinforced concrete, vitrified clay, corrugated metal, and plastic.

Combination System

In many cases, the system is a combination of open and closed drainage (Fig. 10.6c). Typically, the open system is used in unpaved areas with the intent of providing more opportunity for the storm water to infiltrate the pervious

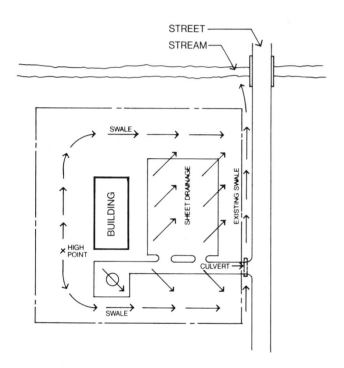

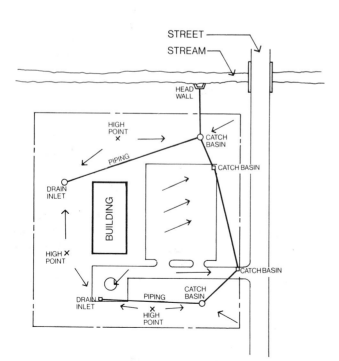

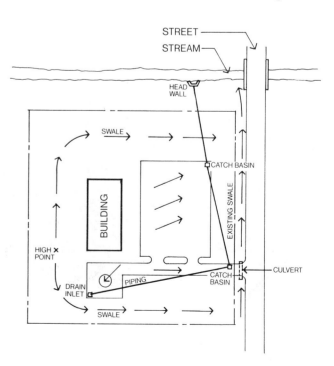

FIGURE 10.6. Surface Drainage Management Systems. a. Open system. b. Closed system. c. Combination system.

surface, whereas the closed system is used in paved areas. The advantages of this system are reduced construction costs compared to those of a totally closed system, lower potential for soil erosion because of reduced volumes of surface runoff, and lower potential for erosion problems at the outfall because of lower volumes in the pipes.

DESIGN AND LAYOUT OF DRAINAGE SYSTEMS

Surface Drainage

The three basic functions of any storm drainage system are to *collect, conduct,* and *dispose* of storm runoff. In order to proceed with the design of a system, each of these functions must be appropriately analyzed as follows.

Collection

To determine where runoff originates, it is necessary to analyze the existing and proposed drainage patterns. To begin with, it is important to analyze off-site patterns, since very few sites exist as isolated entities. The extent of the surrounding drainage area, its effect on the site, and the effect of the proposed site development must be evaluated. The character of the drainage area, including soils, slopes, and surface cover, has to be determined.

The character of the existing and proposed on-site patterns must also be investigated. This includes the location of high points, low points, ridges, valleys, swales, points of concentration, and streams. Also by analyzing the on-site patterns, the critical areas where runoff should be collected or intercepted may be determined. Drainage structures such as catch basins, drain inlets, and area drains and interception measures such as diversions and trench drains must be located at these critical points. By understanding the nature, direction, and extent of these patterns it is possible to estimate the rate of runoff expected from the individual drainage areas. The analysis of where runoff originates also provides an opportunity to reexamine the grading scheme to determine whether it is functioning properly with regard to storm drainage. Remember that to ensure positive drainage all surfaces must slope and that, in most cases, it is illegal to increase or concentrate flow across adjacent property.

Disposal

The question of where the water is going must be answered before the question of how it is going to get there, because the outlet will have a direct influence on the method of conduction. It must first be determined whether the runoff will be disposed of on or off the site and whether the connection is to a natural system (e.g., stream, river, or lake) or an engineered system (e.g., drainage channel, storm sewer, detention/retention basin). Failure to consider the suitability of the outlet and the nature of the connection has, in some cases, produced systems that do not function properly, wasted money, or even resulted in litigation brought about by damage caused by concentrated water flowing from drainage pipes or channels onto adjacent properties. Finally, the connection and means of disposal must comply with all building codes, zoning ordinances, and environmental regulations.

Conduction

Storm runoff may be conducted in an open, closed, or combination system. Some factors that could influence the type, layout, components, and extensiveness of a system have been previously mentioned. These include surface cover and soil type; environmental and ecological consequences; type of land use (e.g., residential development, park, or urban plaza); and visual appearance. This last point is particularly important to landscape architects and site planners. In addition, such rainfall characteristics as frequency, duration, and intensity of storms have to be considered. In most cases it is not difficult to develop a storm drainage management system that functions properly hydraulically. However, developing a system that is integrated with the overall design concept and does not become visually distracting or environmentally damaging requires a certain amount of care, finesse, and understanding of basic engineering and ecological principles. A general rule of thumb is to make the management system as inconspicuous and unobtrusive as possible. This does not mean that costly measures should be taken to disguise or hide a system or that in some designs it is not appropriate to use the management system as an element that is significant to the design concept. It does mean that generally, for open systems, swales should be broad with gentle, rounded cross sections rather than narrow and ditchlike and that retention/detention basins should be integrated into the overall site design rather than located as an afterthought. For closed systems, structures should be placed logically and inlet openings not oversized unnecessarily. Where drainage inlets are placed in pavement, the location should be coordinated with the paving pattern, and the sloping of the pavement to direct runoff should be minimized to prevent a warped surface appearance (Fig. 10.4b).

Regardless of whether an open, closed, or combination system is used, the typical conduction pattern will be treelike in appearance. This pattern results from a hierarchical collection of runoff from a series of small drainage areas which, in most instances, will terminate at one disposal point. It must also be realized that generally the conduction component (swales or pipes) must increase in size as it progresses toward the outlet point, since the volume of runoff continues to increase.

In laying out piping patterns for closed systems, it is important to avoid buildings, subsurface structures such as other utilities, retaining walls, and trees. For most small-scale site development projects, pipes are designed as straight lines, since this reduces the potential for clogging and makes the system easier to clean. It is possible, however, to lay out piping with curved alignments, particularly for large pipes (36-in. diameter and larger). Since pipes are placed underground, they are independent of the surface slope. However, avoid running pipe contrary to the surface slope, since this increases the amount of trench excavation required. Finally, storm water flows through pipes by gravity; therefore, all pipes must *slope*.

APPLICATIONS

Design and Sizing of Grassed Swales (Waterways)

Grassed swales are a common component of open drainage systems. Generally, grassed swales should not carry continuous flows or even be continuously wet. Where this might occur, an alternative method such as use of lined waterways or subsurface drains which leave the swale to carry flow only during and after storms is preferable. When designing the layout of a management system it may be possible to retain an existing natural drainageway or it may be necessary to construct new waterways. Constructed waterways generally have a smooth, shallow, and relatively wide cross section, which is called parabolic, since it represents part of a parabolic curve. Although the discussion here will be limited to parabolic waterways, several other cross-sectional shapes are applicable. The dimensions of the cross section are specified by the width, W, and depth, D, as illustrated in Fig. 10.7. Usually swales are seeded with the same seed mixtures which would be used for lawns in that particular location or with native grasses if a naturalized effect is desired. Newly constructed vegetated waterways should be protected from the erosive effect of flowing water until a good stand of grass has been established.

Grassed waterways are constructed to a design slope which is staked out and controlled during construction by profile leveling. It is preferable to design swales so that the velocity of the water flowing in them will not be decreased, since a reduction in flow velocity will cause siltation. Velocity could be reduced by a change in slope from steep to flat, by enlarging the cross section without also increasing the slope or by increasing the frictional resistance of the surface caused by tall vegetation placed

CROSS-SECTIONAL AREA: $A = \frac{2}{3} WD$

HYDRAULIC RADIUS: $R = \frac{W^2 D}{1.5W^2 + 4D^2}$

TOP WIDTH: $W_2 = W_1 \left(\frac{D_2}{D_1}\right)^{0.5}$

(a)

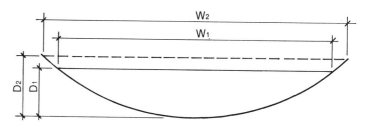

(b)

(c)

(d)

FIGURE 10.7. Parabolic Swales. a. Elements of parabolic swales. b. A parabolic swale flowing full. c. A diversion swale with drain inlet. d. A swale which is not functioning properly. A broader cross section or longer stand of grass could be used to slow the velocity of flow and reduce the potential for erosion.

downslope of a stand of short vegetation. However, in some situations a decrease in velocity which filters out sediment is desired. It must be realized, however, that sedimentation changes the character and capacity of a swale, thus increasing the level of maintenance. Generally, grassed waterways are maintained in the same way as any turf area. Two points of critical concern, though, are the prompt repair of any erosion damage and the removal of accumulated sediment.

The required dimensions for grassed swales may be determined analytically or by a variety of published nomographs or other design aids. To determine the required dimensions of a swale, the rate of runoff to be handled must be known. This may be computed from the Rational formula or by SCS hydrology, as previously discussed. Stable (i.e. non-erodible) channels are usually designed to conform to either the maximum permissible velocity concept or the maximum permissible shear stress concept. The discussion in this text is confined to the former.* With the runoff rate known, the next step is to determine the slope for the proposed swale and the design velocity of flow. The slope is generally determined by the proposed grading plan, the elevation of the outlet point, and the existing topography. The permissible maximum design velocity depends on the type and condition of the vegetation, the erodibility of the soil, and the slope of the swale. Recommended velocities for various conditions may be found in Table 10.1. In situations where excessive velocities cannot be prevented (steep slopes, channel width limitations, etc.), structural linings, such as riprap, gabions, or concrete, should be used. A discussion of the design of such linings is beyond the scope of this text.

Since the friction, or resistance to flow, of the vegetation varies with its length (which is short immediately after mowing and relatively long just before mowing), the range of heights must be determined. Also, as the flow depth increases, long vegetation bends over and offers less resistance than it would with only a shallow depth. For this reason various resistance or retardance factors have been experimentally determined; they are listed in Table 10.2.

If the length of vegetation changes, the final design

---

*Readers interested in swale design by shear stress are referred to: *Stability Design of Grass-lined Open Channels*, Agriculture Handbook No. 667, USDA-Agricultural Research Service or *Design of Roadside Channels with Flexible Linings*, Hydraulic Engineering Circular No. 15, USDOT-Federal Highway Administration.

TABLE 10.1. Permissible Velocities for Vegetated Swales and Channels

Cover	Slope range[2] (*percent*)	Permissible velocity[1] (fps)	
		Erosion-resistant soils[3]	Easily eroded soils[4]
Bermudagrass	<5	8	6
	5–10	7	4
	over 10	6	3
Bahiagrass			
Buffalograss			
Kentucky bluegrass	<5	7	5
Smooth brome	5–10	6	4
Blue grama	over 10	5	3
Tall fescue			
Grass mixture	[2]<	5	4
Reed canarygrass	5–10	4	3
Sericea lespedeza			
Weeping lovegrass			
Yellow bluestem	[5]<5	3.5	2.5
Redtop			
Alfalfa			
Red fescue			
Common lespedeza[6]	[7]<5	3.5	2.5
Sudangrass[6]			

[1]Use velocities exceeding 5 fps only where good covers and proper maintenance can be obtained.
[2]Do not use on slopes steeper than 10 percent except for vegetated side slopes in combination with a stone, concrete, or highly resistant vegetative center section.
[3]Cohesive (clayey) fine-grain soils and coarse-grain soils with cohesive fines with a plasticity index of 10 to 40 (CL, CH, SC, and CG).
[4]Soils that do not meet requirements for erosion-resistant soils.
[5]Do not use on slopes steeper than 5 percent except for vegetated side slopes in combination with a stone, concrete, or highly resistant vegetative center section.
[6]Annuals—use on mild slope or as temporary protection until permanent covers are established.
[7]Use on slopes steeper than 5 percent is not recommended.

TABLE 10.2. Retardance Factors for Grassed Swales

Retardance	Cover	Condition
A	Weeping lovegrass	Excellent stand, tall (average 30 inches)
	Reed canarygrass or Yellow bluestem ischaemum	Excellent stand, tall (average 36 inches)
B	Smooth bromegrass	Good stand, mowed (average 12 to 15 inches)
	Bermudagrass	Good stand, tall (average 12 inches)
	Native grass mixture (little bluestem, blue grama, and other long and short midwest grasses)	Good stand, unmowed
	Tall fescue	Good stand, unmowed (average 18 inches)
	Sericea lespedeza	Good stand, not woody, tall (average 19 inches)
	Grass-legume mixture—Timothy, smooth bromegrass, or orchardgrass	Good stand, uncut (average 20 inches)
	Reed canarygrass	Good stand, uncut (average 12 to 15 inches)
	Tall fescue, with birdsfoot trefoil or ladino clover	Good stand, uncut (average 18 inches)
	Blue grama	Good stand, uncut (average 13 inches)
C	Bahiagrass	Good stand, uncut (6 to 8 inches)
	Bermudagrass	Good stand, mowed (average 6 inches)
	Redtop	Good stand, headed (15 to 20 inches)
	Grass-legume mixture—summer (orchardgrass, redtop, Italian ryegrass, and common lespedeza)	Good stand, uncut (6 to 8 inches)
	Centipedegrass	Very dense cover (average 6 inches)
	Kentucky bluegrass	Good stand, headed (6 to 12 inches)
D	Bermudagrass	Good stand, cut to 2.5-inch height
	Red fescue	Good stand, headed (12 to 18 inches)
	Buffalograss	Good stand, uncut (3 to 6 inches)
	Grass-legume mixture—fall, spring (orchardgrass, redtop, Italian ryegrass, and common lespedeza)	Good stand, uncut (4 to 5 inches)
	Sericea lespedeza or Kentucky bluegrass	Good stand, cut to 2-inch height. Very good stand before cutting
E	Bermudagrass	Good stand, cut to 1.5-inch height
	Bermudagrass	Burned stubble

should always be checked for channel stability with maximum velocity (short vegetation) and capacity with minimum velocity (long vegetation). As a minimum, swales should be designed to carry the peak flow for a 10-yr storm frequency.

Swale Design by Computational Hydraulics

There are two formulas, Manning's equation for open channels (Eq. 9.3) and the continuity equation, which are used in determining the dimensions of open channels, including swales. The *continuity equation* relates the cross-sectional area to the design flow and the average flow velocity. It is defined as:

$$q = AV \quad (10.1)$$

where: q = flow rate, cfs
A = cross-sectional area of flow, ft^2
V = velocity of flow, fps

If the hydraulic radius and the required cross-sectional area are known, the design dimensions can be determined.

Example 10.1

A grassed waterway with a slope of 4% must carry 50 cfs of runoff. The soil is easily eroded. Design a vegetated drainage swale with a parabolic cross section. The vegetative cover will be a good stand of bluegrass sod, which will be kept mowed to a 2-in. height.

Solution. From Table 10.1 the permissible velocity for the given conditions is determined as 5 fps, and the roughness coefficient (Table 9.2) is taken as 0.04. The known values are substituted into Manning's equation:

$$V = \frac{1.486}{n} R^{0.67} S^{0.50}$$

$$5 = \frac{1.486}{0.04} R^{0.67} \, 0.04^{0.50}$$

$$R^{0.67} = \frac{5 \times 0.04}{1.486 \times 0.04^{0.50}}$$

$$R^{0.67} \approx 0.67$$

$$R \approx 0.67^{1/0.67}$$

$$R \approx 0.67^{1.5}$$

$$R \approx 0.55$$

The minimum cross-sectional area is determined by substituting the runoff volume, 50 cfs, and the permissible velocity, 5 fps, into the continuity equation:

$$q = AV$$
$$50 = A \times 5$$
$$A = \frac{50}{5} = 10 \text{ ft}^2$$

At this point, the required cross section for the drainageway is obtained by successive approximations. For small parabolic swales an initial trial depth may be taken as $1.5R$, which in this case is 1.5×0.55, or 0.825 ft. As indicated in Fig. 10.7, the *area* of a parabolic cross section is:

$$A = 2/3 \; WD \qquad (10.2)$$

where: A = cross-sectional area, ft^2
D = depth at the center, ft
W = top width, ft

Substituting into the equation, a trial width of 18.18 ft is obtained as follows:

$$10 \text{ ft}^2 = 2/3 \; W \times 0.825 \text{ ft}$$

$$W = \frac{10 \times 3}{0.825 \times 2} = 18.18 \text{ ft}$$

The *hydraulic radius* of a parabolic channel is expressed as:

$$R = \frac{W^2 D}{1.5 \, W^2 + 4D^2} \qquad (10.3)$$

where: R = hydraulic radius, ft
W = top width, ft
D = depth at the center, ft

For this problem a trial hydraulic radius is obtained by substituting $D = 0.825$ and $W = 18.18$ into the equation.

$$R = \frac{18.18^2 \times 0.825}{(1.5 \times 18.18^2) + (4 \times 0.825^2)}$$

$$R = 0.547 \text{ ft (or approximately 0.55 ft)}$$

Since this agrees with the hydraulic radius for a velocity of 5 fps, the design dimensions $W = 18.18$ ft and $D = 0.825$ ft are satisfactory. In any approximation, the final area should be close to but *not less than* the calculated area (in this case, 10 ft^2) and the final hydraulic radius should be close to but *not greater than* the calculated hydraulic radius (0.55 ft in this problem). Thus the final design dimensions for this example are $W = 18.18$ ft (or 18 ft) and $D = 0.83$ ft (or 10 in.). If the first trial dimensions do not accomplish this, new dimensions must be assumed and tested.

Swale Design with Charts

Since the hydraulic design of swales is somewhat complex and since the n value of vegetation is variable, charts for the solution of Manning's equation have been developed. These charts aid in finding the required hydraulic radii for various vegetative retardance classes and maximum permissible velocities. They are shown in Fig. 10.8. An example will illustrate their use.

Example 10.2

Storm water runoff from part of a townhouse development is to be conducted through a vegetated waterway (swale). The area in which this swale is located will be mowed twice during the growing season so that the height of the vegetation, which will be a good stand of Kentucky bluegrass, will vary between 2 and 10 in. The soil survey report shows that the existing soil is easily eroded. The site plan indicates that the swale will have a 5% gradient and is to be designed to carry 36 cfs.

Solution. The maximum permissible velocity for this swale with Kentucky bluegrass on a 5% gradient and on an easily eroded soil is determined to be 4 fps from Table 10.1. In Table 10.2, the vegetative retardance factor for the cut condition is specified as D and for uncut condition as C. To design for stability, Fig. 10.8d (retardance D) is used. Entering horizontally from the left with 4 fps velocity until the 5% slope line is intersected, a vertical line is extended upward from the point of intersection and the required hydraulic radius is determined as 0.43 ft.

142 Design and Sizing of Storm Water Management Systems

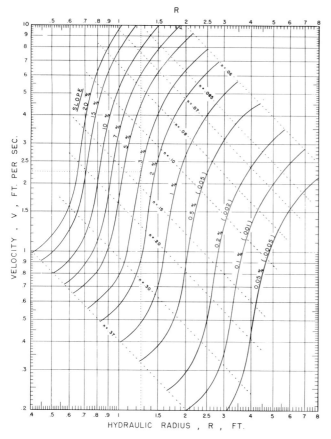
a. Retardance A.

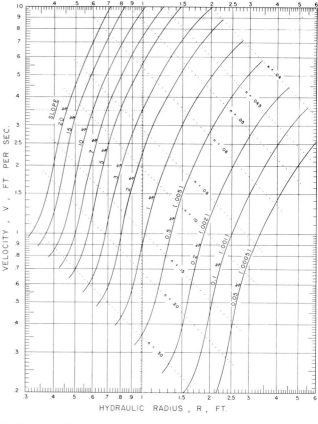
b. Retardance B.

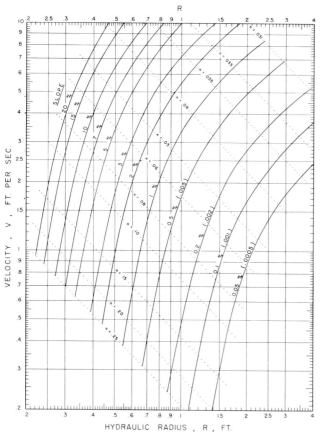
c. Retardance C.

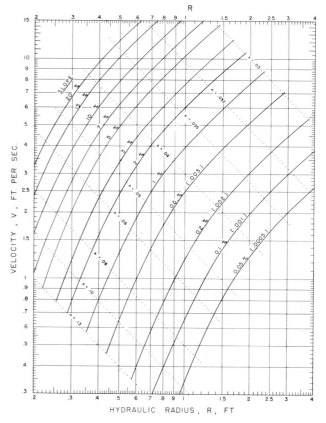
d. Retardance D.

FIGURE 10.8. Solution of Manning's Equation for Swales with Various Vegetative Retardance Factors. (From USDA-SCS.)

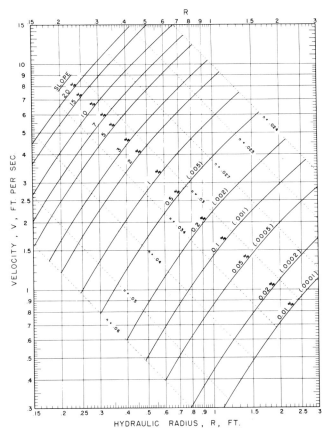

e. Retardance E.

FIGURE 10.8. *continued.*

As in the previous example, the minimum required cross-sectional area is determined from $q = AV$:

$$36 = A \times 4$$
$$A = 36/4 = 9 \text{ ft}^2$$

The trial depth is again $1.5 \times R$, or 1.5×0.43, which is 0.65 ft. From $A = 2/3 WD$, $9 = 2/3 W \times 0.65$. Solving for W:

$$W = \frac{9 \times 3/2}{0.65} = 20.8 \text{ ft}$$

The dimensions $W = 20.8$ ft and $D = 0.65$ ft are satisfactory for stability with the vegetation in the mowed condition.

When the grass is long, retardance factor C is indicated. Since the velocity will be reduced below the allowable 4 fps as a result of the greater retardance, the cross-sectional area will have to be increased, so that the swale will still be able to handle 36 cfs. This is done by deepening and widening the swale, retaining the same parabolic shape. A trial and error procedure must be employed. Assume a flow depth of 0.75 ft. Then the hydraulic radius from $D = 1.5R$ is $D/1.5 = R = 0.5$ ft. Entering Fig. 10.8c (retardance C) vertically with $R = 0.5$ until the 5% slope line is intersected and then extending a horizontal line to the left, a velocity of 3.4 fps is obtained.

From the simplified equation of a parabola ($y = fx^2$ or $D = fW^2$), the width is proportional to the square root of the depth, and thus the new top width is computed as $W_2 = W_1 (D_2/D_1)^{0.5} = 20.8 \times (0.75/0.65)^{0.5} = 22.3$ ft. The new area is $2/3$ $WD = 2/3 \times 22.3 \times 0.75 = 11.15$ ft^2. From $q = AV$, the swale under these conditions will handle $11.15 \times 3.4 = 37.9$ cfs, which is a little more than required and is satisfactory.

It must be pointed out that state or local review agencies may have regulations specifying permissible velocities and retardance classifications for various types of vegetation different from those listed in Tables 10.1 and 10.2. Applicable regulatory agencies should be consulted to determine specific requirements for a particular project.

Critical Velocity

To enhance the stability of vegetated waterways by preventing possible turbulence, some regulatory agencies suggest that design flow velocities be no greater than 90% of the *critical velocity*. Critical velocity (which occurs at the critical depth) is attained when the specific energy of the flowing water is at the minimum. For a given rate (cfs), flows at less than critical depth (with higher velocities) are called *supercritical* and those at depths greater than critical (with lower velocities) are called *subcritical* (Fig. 10.9). A full discussion of the critical flow phenomenon is beyond the scope of this book, and the reader is referred to texts of fluid mechanics or hydraulics. (Of course, flow velocities should never exceed the allowable velocities stipulated in Table 10.1.)

For parabolic waterways, the critical velocity is a function of the flow depth; it can be computed from the following equation:

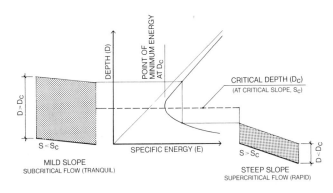

FIGURE 10.9. Critical Flow Depth. Critical flow depth, D_c, occurs at the point of minimum energy of the specific energy curve. Above the minimum energy point there are two depths at which flow can occur with the same specific energy for the same q. One depth is greater than D_c and results in subcritical or tranquil flow; the other is less than D_c and results in supercritical or rapid flow.

TABLE 10.3. Critical Velocities and Hydraulic Radii for Parabolic Waterways

Depth (ft)	Critical[a] Velocity (fps)	90% of Crit. Velocity (fps)	Approx. Hydraulic Radius[b] (ft)
0.5	3.27	2.95	0.33
0.6	3.59	3.23	0.40
0.7	3.87	3.49	0.47
0.8	4.14	3.73	0.53
0.9	4.39	3.95	0.60
1.0	4.63	4.17	0.67
1.1	4.86	4.37	0.73
1.2	5.07	4.56	0.80
1.3	5.28	4.75	0.87
1.4	5.48	4.93	0.93
1.5	5.67	5.10	1.00
1.6	5.86	5.27	1.07
1.7	6.04	5.43	1.13
1.8	6.21	5.59	1.20
1.9	6.38	5.74	1.27
2.0	6.55	5.89	1.33
2.1	6.71	6.04	1.40
2.2	6.87	6.18	1.47
2.3	7.02	6.32	1.53
2.4	7.17	6.46	1.60
2.5	7.32	6.59	1.67

[a] Computed from $V_c = 4.63 D^{0.5}$
[b] Computed from $R = D/1.5$

$$V_c = 4.63 \, D^{0.5} \tag{10.4}$$

where: V_c = critical velocity, fps
D = flow of depth of parabolic channel, ft

Table 10.3 lists values of critical velocities and hydraulic radii for parabolic waterways of various depths. The following examples will illustrate how to design for subcritical flow velocity.

Example 10.3a

A grassed waterway with a slope of 4% must carry 50 cfs of runoff. The soil is easily eroded. The vegetative cover will be a good stand of bluegrass sod, which will be kept mowed to a 2-in. height. Design a vegetated drainageway with a parabolic cross section. The regulatory agency limits the design flow velocity to a maximum of 90% of the critical velocity. (Note that, except for the critical velocity requirement, this is the same as in Example 10.1.)

Solution. For a good stand of vegetation 2 in. high, the retardance factor for allowable velocity is determined as D from Table 10.2. The swale design must agree with both Table 10.3 and Fig. 10.8d. From the solution for Example 10.1 it is known that the velocity corresponding to a 0.825-ft depth is 5 fps. According to Table 10.3, this is too great to meet the critical velocity requirement. Therefore, a trial depth of 0.7 ft with 0.90 V_c of 3.49 fps and R of 0.47 ft is selected. Entering Fig. 10.8d vertically with $R = 0.47$ until the "4% slope line" (between 3%

and 5%) is intersected, a horizontal line is extended from the point of intersection to the left and a velocity of about 3.7 fps is obtained. This is more than the 3.49 fps indicated in Table 10.3 as 0.90 V_c for a depth of 0.7, and D and R must be reduced.

A depth of 0.6 with a corresponding R of 0.40 is tried next, resulting in a velocity of about 2.8 fps from Fig. 10.8d. This is less than the 3.23 fps indicated as 0.90 V_c and would meet the velocity restriction. However, by deepening the channel somewhat, the width, which would be required to carry the design flow rate with a depth of 0.6 ft, can be reduced.

For a depth of 0.65, $R = 0.65/1.5 = 0.43$, $V_c = 4.63 \times D^{0.5} = 4.63 \times 0.65^{0.5} = 3.73$, and $0.90 \, V_c = 3.36$. (These values could also have been determined from Table 10.3 by interpolation.) From Fig. 10.8d, $V = 3.30$ for $R = 0.43$ at 4% slope. This is less than 90% of the critical velocity. The n value from the same figure is 0.051.

From $q = AV$

$$50 = A \times 3.30$$

$$A = \frac{50}{3.30} = 15.2 \text{ ft}^2$$

From $A = 2/3 \, WD$

$$15.2 = 2/3 \, W \times 0.65$$

$$W = \frac{15.2 \times 3}{0.65 \times 2} = 35.1 \text{ ft}$$

The design dimensions are:

$D = 0.65$ ft

$W = 35.1$

Note that in this case the width necessary to reduce the velocity to subcritical is almost twice that of the previous design for allowable velocity (Example 10.1).

As a check, R, V, and q should be computed from the design dimensions.

$$R = \frac{W^2 D}{1.5 W^2 + 4 D^2} = \frac{35.1^2 \times 0.65}{1.5 \times 35.1^2 + 4 \times 0.65^2} = 0.43$$

$$V = \frac{1.486}{n} R^{0.67} S^{0.5} = \frac{1.486}{0.051} \times 0.43^{0.67} \times 0.04^{0.5} = 3.31$$

$$q = AV = 15.2 \times 3.31 = 50.3 \text{ cfs}$$

Thus all requirements have been met.

If conditions permit, other methods could be used to reduce flow velocities, such as increasing the retardance factor by letting the vegetation grow to a greater height and/or reducing the gradient. This will be demonstrated in the next example.

Example 10.3b

It has been decided that the swale of Example 10.3a requires too great an area and it will therefore be relocated on the site plan. To decrease the width and cross-sectional area, the flow velocity will be increased to near the permissible velocity of 5 fps while remaining 90% of critical velocity or less. The vegetation is to remain retardance D. What will be the dimensions and the required slope for this swale?

Solution. Table 10.3 indicates that a parabolic swale with a depth of 1.4 ft has a hydraulic radius of about 0.93 ft with 90% of critical velocity 4.93 fps. The intersection of $R = 0.93$ and $V = 4.93$ on Fig. 10.8d indicates a slope of 1.67% and an n value of 0.036 (by interpolation).

$$q = AV$$

$$50 = A \times 4.93$$

$$A = \frac{50}{4.93} = 10.14 \approx 10.1 \text{ ft}^2$$

$$A = 2/3 \; WD$$

$$10.1 = 2/3 \; W \times 1.4$$

$$W = \frac{10.1 \times 3}{1.4 \times 2} = 10.82 \approx 10.8 \text{ ft}$$

Thus the trial design dimensions are:

$D = 1.4$ ft

$W = 10.8$ ft

and the trial design slope is 1.67%.

When these values are checked by the method shown previously, the resulting velocity slightly exceeds 4.93 fps, which is 90% of the critical velocity. Recalculating with $D = 1.34$ ft (90% $V_c = 4.8$ fps) and checking again lead to values which are within acceptable limits. The final design measurements are:

$D = 1.34$ ft

$W = 11.9$ ft

$S = 1.6\%$

Where it is appropriate, increasing the retardance of the vegetation may be more practical than reducing the slope by relocating the swale.

Some designers increase the depth by 0.3 to 0.5 ft after the hydraulic design procedure. This will also widen the channel. The increased depth is called *freeboard*.

Designing and Sizing Pipe Systems

It is often more practical to dispose of excess surface water by means of subsurface piping, or closed systems, rather than by open drainage channels. As previously mentioned, in designing any drainage system the first step is to determine the availability of an adequate outlet. The next step for a closed system is to determine surface slopes and configuration and location of collection points where the inlet structures will be placed for the piping system. Generally it is best not to locate these structures near trees, main walks, or buildings, since occasional clogging may cause flooding. The collection points are then connected on the drawing, usually by straight lines, which represent the subsurface pipes. The network should be designed with the minimum adequate amount of pipe for economy. Where two or more pipes join, or where pipes join at different elevations, structures such as manholes, junction boxes, or catch basins should be used. A recommended minimum depth for pipes is 3 ft to protect them from being crushed by traffic and, in northern climates, to reduce potential frost problems.

Most often using the Rational method, the peak rate of runoff (q_p) is calculated for the drainage area collected by each pipe, keeping in mind that the volume is cumulative proceeding downgrade as additional inlets are connected to the system. By selecting a slope for the pipe, a pipe size for the calculated runoff can be determined by Manning's equation and the continuity equation. This process is illustrated in the following example.

Example 10.4

Calculate the pipe size required to carry 10 cfs at a 10% slope. Use vitrified clay pipe with an n value of 0.015.

Solution. The cross-sectional area of a circular pipe flowing full is πr^2, the area of a circle. The wetted perimeter of a circular pipe flowing full is equal to the circumference or $2\pi r$. Thus the hydraulic radius is:

$$R = \frac{\pi r^2}{2\pi r} = \frac{r}{2}$$

where: R = hydraulic radius, ft
r = inside radius of pipe section, ft

As stated before, Manning's equation is:

$$V = \frac{1.486}{n} R^{0.67} S^{0.5}$$

and the continuity equation is $q = AV$. These two equations can be combined as follows:

$$q = A \frac{1.486}{n} R^{0.67} S^{0.50}$$

By substituting the known values for the problem, the equation becomes:

$$10 = \pi r^2 \frac{1.486}{0.015} \left(\frac{r}{2}\right)^{0.67} (0.1)^{0.50}$$

$$r^2 \times r^{0.67} = \frac{10 \times 0.015 \times 2^{0.67}}{\pi \times 1.486 \times 0.1^{0.50}}$$

$$r^2 \times r^{0.67} = 0.16$$

$$r^{2.67} = 0.16$$

$$r = 0.16^{1/2.67} = 0.50 \text{ ft}$$

Therefore, a pipe with a 0.50-ft radius, or 1.0-ft diameter, is an adequate size for this problem.

In most cases, several pipes are interconnected to create a closed drainage system. A procedure for sizing pipes for such a system is demonstrated in Examples 10.5 and 10.6. In Example 10.6 a method for determining rates of runoff from various drainage areas which contribute to a piping system is discussed.

Example 10.5

Figure 10.10 schematically illustrates two drainage areas and a proposed drainage system consisting of two catch basins and required piping. The characteristics of the drainage areas are:

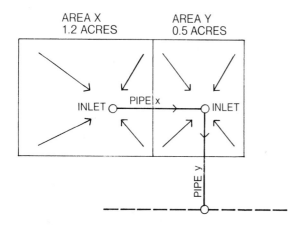

FIGURE 10.10. Schematic Plan for Example 10.5.

Drainage Area X: $A = 1.2$ ac
$C = 0.3$
$T_c = 45$ min

Drainage Area Y: $A = 0.5$ ac
$C = 0.7$
$T_c = 20$ min

Determine the peak rate of runoff which must be handled by pipes x and y for a 10-yr storm frequency.

Solution. The first step is to determine the rainfall intensity from Fig. 8.2 for each of the drainage areas. This is 2.5 iph for area X and 4.0 iph for area Y. Using the Rational formula, the rate of runoff for each drainage area can be calculated.

$$q_x = 0.3 \times 2.5 \times 1.2$$
$$q_x = 0.9 \text{ cfs}$$
$$q_y = 0.7 \times 4.0 \times 0.5$$
$$q_y = 1.4 \text{ cfs}$$

Pipe x must handle only the runoff from drainage area X; therefore, it can be sized for 0.9 cfs. However, pipe y must handle the runoff from area Y plus the water flowing in pipe x. As determined previously, the peak rate of flow for area Y is 1.4 cfs. This rate is reached in 20 min, but at that time pipe x is still not flowing full, since area X does not peak until 45 min after the beginning of the design storm. If q_x and q_y were simply added, the pipe might be oversized. Using the definition of *time of concentration* as the time for water to flow from the hydraulically most remote point of the drainage area to the point of interest, this would be 45 min (T_c for area X) plus the time of flow in the pipe from area X to area Y (for distances less than 150 ft, this time is usually negligible). The rate of runoff for the longer time of concentration is:

$$q_y = 0.7 \times 2.5 \times 0.5$$
$$= 0.88 \text{ cfs} \approx 0.9 \text{ cfs}$$

Therefore, the rate of flow which pipe y must handle is $0.9 + 0.9$, approximately 1.8 cfs. This is above the peak rate of 1.4 cfs, which occurs in 20 min for area Y, but below 2.3 cfs, which would have resulted if the two rates had been added.

A more accurate and sophisticated procedure is to use the maximum flow rates obtained from the MRM.

For a storm duration of 20 min the maximum flow rate for area X is $q = C \times C_A \times i \times A \times \text{DUR}/T_c$, since 20 min is less than the time of concentration of 45 min. Therefore, $q = 0.3 \times 1.0 \times 4.0 \times 1.2 \times 20/45 = 0.64 \approx 0.6$ cfs. The total runoff rate for pipe y is $0.6 + 1.4 = 2.0$ cfs. This is more than the 1.8 cfs previously determined and should be used for design.

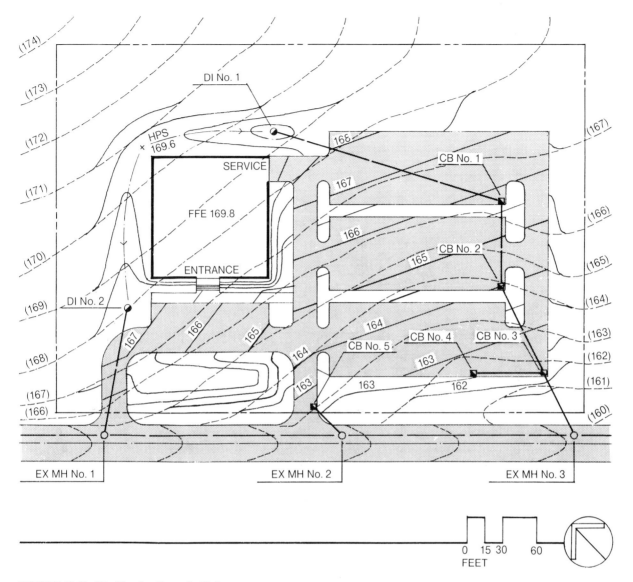

FIGURE 10.11. Site Plan for Example 10.6.

It is important to analyze each project carefully to make sure that pipes are designed to handle peak flows regardless of the time of concentration upstream.

Example 10.6

In this problem the site plan (Fig. 10.11) for a small office building indicates proposed grades, location of drainage structures, and piping pattern. For this site plan all parking areas, drives, and streets have 6-in.-high curbs and all areas not paved are lawn with a silt loam soil texture. The off-site drainage area consists of 2 ac of woodland. The longest overland distance for this runoff, which enters the site at the north corner, is 200 ft with an average slope of 4% and, again, a silt loam soil texture. Use a 10-yr storm frequency for the design of the drainage system, which is located in New Jersey. Runoff from the building roof must be accommodated in the design.

Solution. Step 1. The first step is to develop an orderly procedure for recording and organizing all information necessary for sizing and laying out a closed drainage system. This required information is arranged in tabular form as illustrated in Table 10.4.

Step 2. The next step is to determine the extent of the drainage areas, the surface characteristics, and the area of each surface type for each drainage structure (Fig. 10.12). The selected surface coefficients (C) for each surface type and the respective areas (A) are recorded in the table. In this problem the assumption is that one-half of the off-site drainage area (or 1 ac) is directed toward drain inlet no. 1 and the other half toward drain inlet no. 2. The same assumption is made for the roof runoff.

Step 3. The third step is to determine the overland flow time (time of concentration) for each of the drainage areas

148 Design and Sizing of Storm Water Management Systems

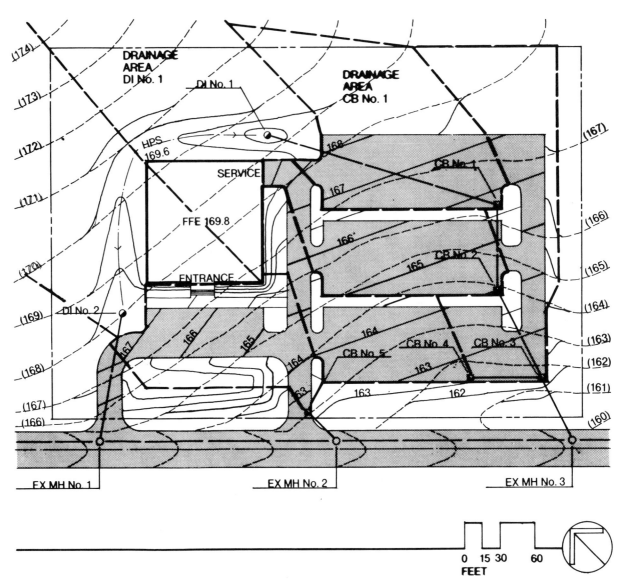

FIGURE 10.12. Drainage Areas for Example 10.6.

by using the nomograph in Fig. 8.3. An adjusted rainfall intensity value (*i*) can then be obtained from the rainfall intensity curves in Fig. 8.2. A 10-yr frequency curve for central New Jersey is used. The information from steps 2 and 3 can be substituted into the Rational formula, $q = CiA$, to calculate the peak rate of runoff to be carried in each pipe. Remember that these rates are additive moving downgrade. As an example, the pipe between catch basin no. 1 and catch basin no. 2 must carry the runoff from the catch basin no. 1 drainage area as well as the runoff from the drainage inlet no. 1 and one-half of the roof runoff.

Step 4. At this point the pipes may be sized. However, rather than using equations to calculate the required sizes, the nomograph in Fig. 10.13 is used. There are five components to the nomograph: discharge (cfs), diameter of pipe (in.), roughness coefficient, velocity (fps), and slope. The figure is actually two nomographs. The first directly relates discharge, diameter of pipe, and velocity. Knowing any two values allows the determination of the third. The other nomograph relates slope and roughness coefficient to diameter of pipe and discharge. Two values on the *same* side of the pivot line must be known (or selected) to determine the other two values. The use of the nomograph is demonstrated by sizing the pipe from drain inlet no. 1 to catch basin no. 1.

Two issues must be discussed before proceeding. The first is the pipe size. To reduce clogging and maintenance problems, a 12-in.-diameter pipe is recommended as a minimum size for landscape applications. This minimum does not apply to roof drain and area drain conditions. There are standard pipe sizes for the different pipe materials and manufacturers' catalogs should be consulted. The second issue is velocity. A recommended range for the velocity of water in pipes is 2.5 to 10 fps. A minimum of 2.5 is used to ensure self-cleaning, while a maximum of 10 is suggested to reduce potential scouring and pipe wear which may occur above this velocity. Also it can be seen from the nomograph that there is an inverse relationship between pipe size and slope and velocity.

For drainage inlet no. 1 to catch basin no. 1, the dis-

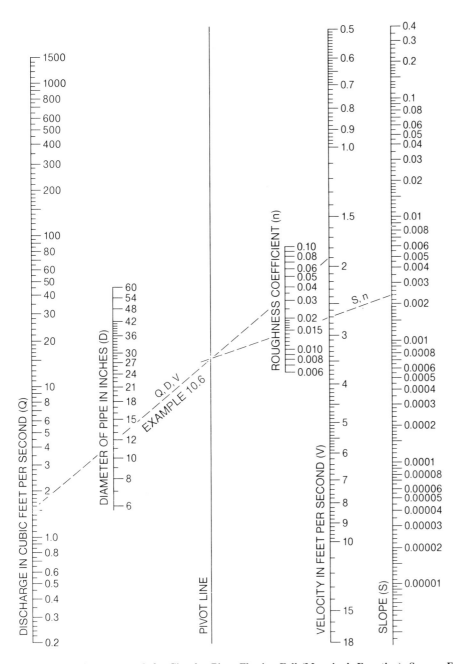

FIGURE 10.13. Nomograph for Circular Pipes Flowing Full (Manning's Equation). Source: From American Iron and Steel Institute 1980.

charge flowing in the pipe is 1.53 cfs; experience indicates that this is a low flow rate. Therefore, the use of a minimum pipe size is anticipated. In applying the nomograph, the discharge (q) is known and a 12-in.-diameter pipe size is selected. The procedure for determining the remaining values is as follows:

1. Locate 1.53 cfs on the discharge line.
2. Locate 12 in. on the diameter of pipe line.
3. Draw a straight line through the two points, crossing the pivot line, until it intersects the velocity line.
4. Read the value of the velocity line at the point of intersection. In this case, the velocity is 1.9 fps, somewhat below the recommended minimum. However, rather than reduce the pipe to 8-in.-diameter (the next lower standard size for reinforced concrete pipe), the 12-in.-diameter is maintained.
5. Next, return to the intersection of the line drawn in (3) with the pivot line.
6. Locate 0.015 on the roughness coefficient line. This is the selected n-value for the piping material (reinforced concrete pipe) used in this problem. (See Table 9.2.)
7. Draw a straight line from the point on the pivot line through 0.015 on the roughness coefficient line until it intersects the slope line.
8. Read the value on the slope line at the point of intersection, which is 0.0024.
9. Record all appropriate values in Table 10.4.

The same procedure is applied to the remaining pipes. All data for the problem are recorded in Table 10.4.

TABLE 10.4. Data for Example 10.6

Area	to	c	$T_c{}^a$ (min)	i (iph)	A (acres)	q_{sub} (cfs)	q (cfs)	D (in.)	S (ft/ft)	L (ft)	V (fps)	n	INV in	INV out	TF[c]
½ Roof	DI No. 1	0.92	37	2.8	0.12	0.31									
DI No. 1	CB No. 1	0.30		2.8	1.45	1.22	1.53	12	0.0024	210	1.9	0.015		156.79	167.60
CB No. 1	CB No. 2	0.30	37	2.8	0.20	0.17									
		0.90		2.8	0.21	0.53	2.23	12	0.0055	75	2.8	0.015	156.29	156.29	165.70
CB No. 2	CB No. 3	0.90	37	2.8	0.30	0.76	2.99	12	0.0095	80	3.7	0.015	155.88	155.88	163.90
CB No. 4	CB No. 3	0.30	10[b]	5.8	0.04	0.07									
		0.90		5.8	0.25	1.31	1.38	8	0.018	50	3.9	0.015		156.02	162.60
CB No. 3	MH No. 3	0.30	37	2.8	0.07	0.06									
		0.90		2.8	0.28	0.71	5.14	15	0.0095	55	4.3	0.015	155.12	154.87	162.00
CB No. 5	MH No. 2	0.30	10	5.8	0.12	0.21									
		0.90		5.8	0.18	0.94	1.15	8	0.011	30	3.2	0.015		157.06	162.50
½ Roof	DI No. 2	0.92	40	2.7	0.12	0.30									
DI No. 2	MH No. 1	0.30		2.7	1.36	1.10	1.40	8	0.018	105	3.9	0.015		161.16	167.20

[a] Flow times in pipes between drainage structures have not been included, since the distances between structures are relatively short and the resultant flow times are negligible. However, for longer distances the flow time in pipes must be included in the time of concentration. (To determine the travel time in pipes, the length of the pipe is divided by the flow velocity. For example, the travel time in a 500-ft-long pipe with a 2.5 fps flow velocity is 500 ft/2.5 fps = 200 sec or 3.33 min.)

[b] Since it takes several minutes for rain to wet a surface thoroughly, many municipalities permit the use of minimum times of concentration, such as 10 or 15 minutes. This will reduce the intensity used for the computation of the runoff rate and thus the required pipe size.

[c] Top of frame.

Two additional points with regard to the design of pipe systems must be made here. The first is that pipe size should *never decrease* proceeding downgrade. The outlet pipe from a drainage structure should be equal to or larger in diameter than the inlet pipe. Note that the pipe size at catch basin no. 4 can be less, since it is a branch line. The second is that, to prevent silting and clogging, it is preferable *not* to decrease pipe flow velocity proceeding downgrade. This is one reason that the 8-in.-diameter pipe was not used at drainage inlet no. 1. The velocity for an 8-in. pipe would have been above 4 fps, which would have been greater than the subsequent velocities. However, 8-in.-diameter reinforced concrete pipe is used for catch basins nos. 4 and 5. The smaller diameter was necessary to ensure a reasonable velocity because of the very low flow rate.

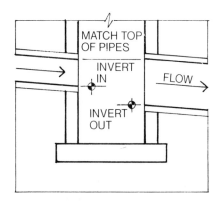

FIGURE 10.14. Invert Elevation. The invert elevation for a drainage structure is the lowest point of the internal cross section of the entering and exiting pipes. The invert of the entering pipe is referred to as the invert *in;* for the exiting pipe it is referred to as the invert *out*. The invert in must not be lower than the invert out. This relationship can be assured by matching the top of pipe elevations.

Step 5. The last step is to determine the invert elevations for the pipes and draw a profile of the system. *Invert elevation* is the elevation of the bottom of the pipe *opening* in a drainage structure as illustrated in Fig. 10.14. The invert elevation of the outlet pipe must be equal to or lower than the invert elevation of the inlet pipe. One technique used to ensure this relationship is to match top of pipe elevations.

Invert elevations are determined by proceeding backward from the point at which the proposed system connects to the existing system. The difference in elevation between the existing invert elevations and the proposed invert elevation is calculated by multiplying the pipe slope by the pipe length. To calculate the outlet invert for catch basin no. 3 (CB no. 3), work back from the existing invert elevation of manhole no. 3 (MH no. 3), which is 154.35.

MH no. 3 invert elevation = 154.35 ft

Pipe slope = 0.0095 ft/ft

Pipe length = 55.0 ft

0.0095 × 55.0 = 0.52 ft

154.35 + 0.52 = 154.87 ft outlet invert elevation at CB no. 3

Once this invert elevation is determined, the remaining elevations may be calculated by the same technique.

After all of the invert elevations have been calculated, a profile may be constructed for the entire system. (Fig. 10.15) A *profile* is a longitudinal section usually taken along the center line of a linear project such as storm drainage piping or roads. A profile is a good checking device to evaluate whether pipes slope properly and whether invert relationships are correct. It also provides an opportunity to analyze whether pipes are placed too deep, thus requir-

Applications 151

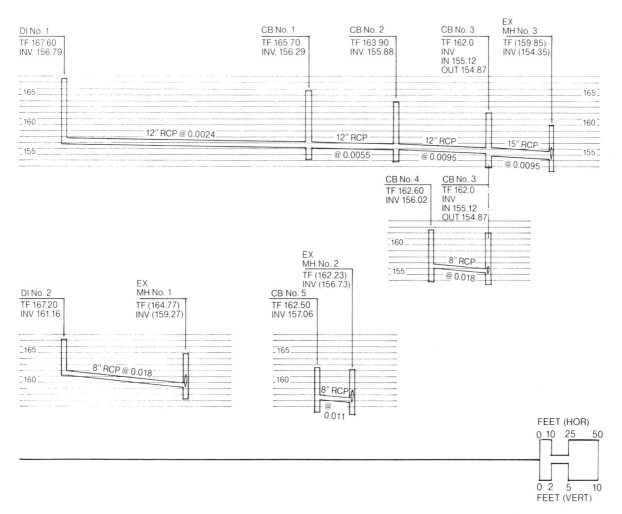

FIGURE 10.15. Profiles of Storm Drainage System for Example 10.6.

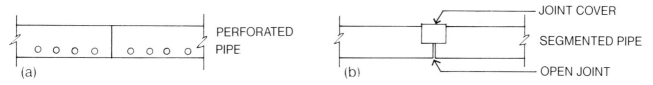

FIGURE 10.16. Pipe installation for subsurface drainage. a. Perforated pipe with perforations placed toward the bottom of the pipe. b. For segmented pipe, such as clay tile, a gap is left between pipe segments. A cover or filter is placed over the open joint to prevent sediment from entering the pipe.

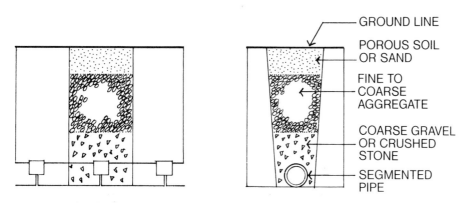

FIGURE 10.17. French Drain with Subsurface Drainage. A French drain is a trench filled with porous material which is used to collect and conduct surface runoff. French drains may also be used in conjunction with subsurface drainage as illustrated.

ing excessive trenching, or too shallow, producing insufficient cover and protection for the pipe. If either case exists, the system must be reevaluated and redesigned.

SUBSURFACE DRAINAGE

The purpose of subsurface drainage is to maintain the water table at a level that provides desirable plant growth conditions and increases the usability of areas for recreational or other purposes. Subsurface drains remove only excess water and not water plants can use. Water available to plants is held in the soil by capillary, or surface tension, forces, whereas excess water flows by gravity into the drains. Subsurface drainage is accomplished by means of clay tile or perforated or porous pipe laid in a continuous line at a specified depth and grade (Figs. 10.16 and 10.17). Free water enters the drains through the joints, perforations, or porous walls of the pipe and flows out by gravity. Although somewhat similar to a closed system, it must be emphasized that water percolates through the soil and is then removed by drains placed below the ground surface.

The major components of a subsurface drainage systems are mains, submains, laterals, and drainage outlets. The laterals collect the free water from the soil and carry it to the submains and mains. These, in turn, conduct the water to the drainage outlet. The installation procedure for perforated pipe is to place the perforations at the bottom of the pipe to minimize the amount of soil particles entering the pipe. For clay tile a small gap is left between pipe segments. Tar paper is placed over the top of the gaps or permeable filter strips are wrapped around the pipe to minimize the entrance of silt particles into the pipe. Finally, under certain conditions, surface inlets may be used in conjunction with a subsurface drainage system.

Underdrainage is a specific type of subsurface drainage used to maintain proper structural conditions. Examples include footing and foundation drains and lateral drains placed behind retaining walls (see Chapter 5). Pavements placed over clay soils or rock or at the bottom of steep slopes may need underdrains to reduce the possibility of hydrostatic pressure caused by trapped water (Fig. 10.18).

Designing and Sizing Subsurface Systems

Subsurface systems may be laid out either to collect water from poorly drained, wet areas or to drain complete areas.

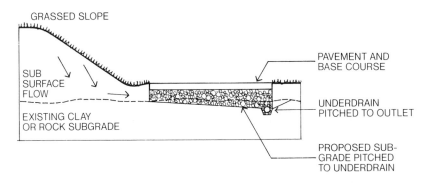

FIGURE 10.18. Underdrainage. The section illustrates a condition in which a buildup of water could occur beneath a pavement, thus causing damage by pressure or frost action. To reduce these problems, the new subgrade is sloped to an underdrain, which carries off the excess water.

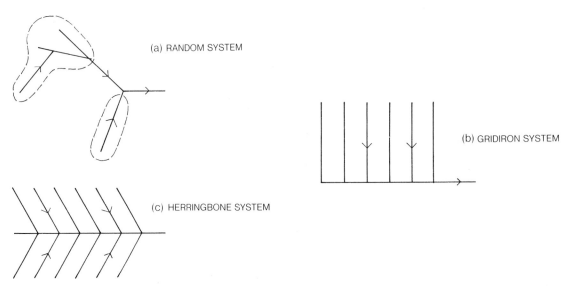

FIGURE 10.19. Piping Patterns for Subsurface Drainage. a. Random system. b. Gridiron system. c. Herringbone system.

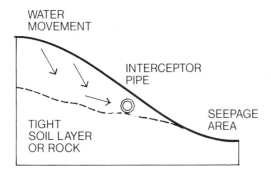

FIGURE 10.20. Cut-Off Drain.

The piping pattern for the former condition is usually random, whereas gridiron or herringbone arrangements are typical for the latter (Fig. 10.19). *Cut-off* drains are pipes placed across a slope to intercept water that would otherwise be forced to the surface by an outcropping of an impermeable layer such as a tight subsoil (Fig. 10.20).

The pipe size required to drain a certain acreage depends on the pipe gradient, since an increase in gradient results in a greater velocity of flow and permits the pipe to drain a larger area. Pipes are placed at constant gradients, or variable gradients with the gradient increasing toward the outlet. The gradients should never decrease, since the velocity of flow would decrease and silt would be deposited in the pipe. Typically pipe gradients for lawn areas vary from a minimum of 0.1% to a maximum of 1.0%. Tables 10.5 and 10.6 show the relationships of pipe size, gradient, and maximum acreage drained for smooth clay or concrete drainage lines and corrugated plastic tubing. These tables were computed by Manning's equation and the continuity equation with n values of 0.011 for the clay and concrete pipe and 0.016 for the corrugated plastic tubing. A drainage coefficient (DC) of 3/8 in. (or 0.0312 ft) was used. *Drainage coefficient* is defined as the depth of water removed over the drainage area in 24 hr. In humid areas of the United States a DC of 3/8 in. is normally used for mineral soils. For organic soils, the acreages of the tables should be reduced by one-half. This may also be done for mineral soils if more rapid drainage is desired.

Depth and Spacing

The depth at which drainage lines are installed generally depends on the outlet conditions. However, there should be a *minimum* of 2 ft of cover in mineral soils and 2.5 ft in organic soils. Drainage lines should always be deep enough to prevent possible frost damage.

The spacing of drainage lines depends on the texture of the soil to be drained. Sandy soils permit more rapid movement of water than do heavy clay soils, and therefore lines may be spaced farther apart and deeper in sandy soils than in clay soils. If drains are spaced too far apart, the central portion between lines will remain poorly drained. Suggestions for depth and spacing of drainage lines are given in Table 10.7.

To design subsurface drainage systems for some soils, such as organic soils and fine sandy loams, a qualified drainage engineer should be consulted, since special precautions must be taken.

Drainage Area

In order to determine the size of pipe, the acreage that each line has to drain must be known. For a gridiron or herringbone system, the area drained by each line may be computed by multiplying the length of the individual lines by the spacing between lines. Where surface inlets are connected to the subsurface system, the total area graded toward the inlets must also be included. The following examples demonstrate the procedure for determining drainage areas and pipe sizes.

TABLE 10.5. Maximum Acreage[a] Drained by Various Pipe Sizes: Clay or Concrete Pipe ($n = 0.011$, DC = 3/8 in./24 hr)

Pipe size (inches)	Slope (%)									
	0.1	0.2	0.3	0.4	0.5	0.6	0.7	0.8	0.9	1.0
4	4.51	6.38	7.82	9.03	10.1	11.1	11.9	12.8	13.5	14.3
5	8.19	11.6	14.2	16.4	18.3	20.0	21.7	23.2	24.6	25.9
6	13.3	18.8	23.1	26.6	29.8	32.6	35.2	37.6	39.9	42.1
8	28.7	40.5	49.6	57.3	64.1	70.2	75.8	81.1	86.0	90.6
10	52.0	73.5	90.0	104	116	127	138	147	156	164
12	84.5	120	146	169	189	207	224	239	254	267

[a] Reduce these acreages by one-half for a 3/4 in. D.C.

TABLE 10.6. Maximum Acreage[a] Drained by Various Pipe Sizes: Corrugated Plastic Tubing ($n = 0.016$, DC = 3/8 in./24 hr.)

Tubing size (inches)	Slope (%)									
	0.1	0.2	0.3	0.4	0.5	0.6	0.7	0.8	0.9	1.0
4	3.10	4.39	5.38	6.21	6.94	7.60	8.21	8.78	9.31	9.81
5	5.62	7.96	9.75	11.3	12.6	13.8	14.9	15.9	16.9	17.8
6	9.15	12.9	15.8	18.3	20.5	22.4	24.2	25.9	27.5	28.9
8	19.7	27.9	34.1	39.4	44.1	48.3	52.1	55.7	59.1	62.3
10	35.7	50.5	61.9	71.5	79.9	87.5	94.5	101	107	113
12	58.1	82.2	101	116	130	142	154	164	174	184

[a] Reduce these acreages by one-half for a 3/4 in. D.C.

TABLE 10.7. Typical Depths and Spacings of Drainage Lines for Various Soil Textures

Texture	Spacing (ft)	Depth (ft)
Clay, clay loam	30–70	2.5–3.0
Silt loam	60–100	3.0–4.0
Sandy loam	100–300	3.5–4.5
Organic Soils	80–200	3.5–4.5

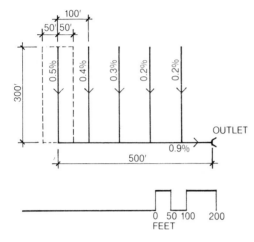

FIGURE 10.21. Piping Plan for Example 10.7.

Example 10.7

A plan for a gridiron drainage system with proposed pipe gradients indicated is illustrated in Fig. 10.21. Determine the pipe sizes for the various parts of the system for a mineral soil and clay pipe. There are no surface inlets.

Solution. The laterals are 300 ft long and spaced 100 ft apart. This means that each line drains 50 ft on either side. Therefore, the drainage area for each lateral is 300 ft by 100 ft (50 + 50), which equals 30,000 ft², or approximately 0.69 ac. On the basis of Table 10.5, a 4-in. pipe is sufficient for all laterals. The main line at the outlet must accommodate the flow from the five laterals (or approximately 3.44 ac) plus its own drainage area of 500 ft by 50 ft, which equals 25,000 ft², or about 0.57 ac, since it provides drainage on one side. The total drainage area of the system is about 4.0 ac (3.44 + 0.57). Table 10.5 shows that a 4-in. pipe is also sufficient for the main line. It might be preferable to increase the pipe size to 5 in., particularly if silting is anticipated.

In certain situations it may be necessary to increase the size of the drainage line, particularly a main line, as it proceeds toward the outlet. This is referred to as a tapered line. Thus it may start as a 5-in. line and increase to 6 in. and perhaps to 8 in. as greater quantities of flow must be accommodated. This condition is illustrated in the following example.

Example 10.8

Figure 10.22 shows a plan for a herringbone system with the spacing indicated. All laterals have a 0.2% gradient and the main slopes at 0.3%. Determine the pipe sizes required for a mineral soil, using clay or concrete pipe.

Solution. There are 16 laterals, each of which drains an area 500 ft by 100 ft, or 50,000 ft², which is approximately 1.15 ac. Since a 4-in. pipe at 0.2% grade can drain 6.38 A according to Table 10.5, all laterals may be 4 in. in size.

The main line has a 0.3% grade. From Table 10.5 the maximum acreage that a 4-in. pipe at 0.3% can drain is 7.82 ac. The number of laterals that can be accommodated by a 4-in. main line is determined by dividing the allowable area by the area for each lateral.

7.82 ac ÷ 1.15 ac/lateral = 6 laterals (6.80)

This means that the main line from junction H to junction E can be 4-in. pipe. At junction E two more laterals are added and the drainage area at this point exceeds 9 ac.

Again from Table 10.5 a 5-in. pipe at 0.3% can drain an area of 14.2 ac. This means that a 5-in. main line can accommodate 12 laterals.

14.2 ac ÷ 1.15 ac/lateral = 12 laterals (12.35)

From junction E to junction B a 5-in. pipe is used, since at junction B laterals nos. 13 and 14 are added and they increase the drainage area beyond the allowable 14.2 ac. At a gradient of 0.3%, a 6-in. pipe can drain 23.1 ac, which is more than the entire area of the proposed system. Therefore, a 6-in. pipe may be used from junction B to the outlet. Again, to reduce the potential for problems, it would be preferable to use 5-in. pipe for all laterals and 6-in. pipe for the entire main line. Various types of pipe junctions such as couplings, reducing couplings, tees, reducing tees, end caps, 45° and 90° ells, and "Ys", are available to complete drainage systems.

Example 10.9

If the herringbone system of Example 10.8 is to be in-

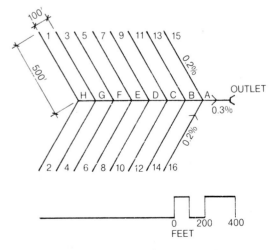

FIGURE 10.22. Piping Plan for Example 10.8.

stalled in an area that requires a drainage coefficient of 3/4 in., determine the drain sizes, using corrugated plastic tubing.

Solution. As previously determined, each lateral drains about 1.15 ac. However, to use the drainage tables with a 3/4-in. DC, all acreages therein have to be reduced by one-half. Therefore, 2.30 ac (2 × 1.15), which when reduced one-half will yield 1.15 ac, must be used to obtain the size for each lateral. Table 10.6 shows that 4-in. tubing is sufficient for all laterals, since it can drain up to 4.39 ac with a 3/8-in. DC and 2.20 ac with a 3/4-in. DC.

At 0.3% slope, Table 10.6 shows the following maximum acreages for various tubing sizes:

Tubing Size, in.	Drainage Coefficient	
	3/8 in. Maximum Acreage	3/4 in. Maximum Acreage
4	5.38	2.69
5	9.75	4.88
6	15.8	7.9
8	34.1	17.1
10	61.9	31.0

The number of laterals that each pipe size can accommodate is:

4 in.   2.69/1.15 = 2.3 or 2

5 in.   4.88/1.15 = 4.2 or 4

6 in.   7.90/1.15 = 6.9 or 6

8 in.   17.1/1.15 = 14.9 or 14

10 in. 31.0/1.15 = 26.95 or 26

Therefore, the minimum tubing sizes for the design of the main are:

Junction	Tubing Size, in.
H to G	4
G to F	5
F to E	6
E to A	8
A to outlet	10

The system can be simplified by using 6-in. tubing from H to E.

Outlets

Subsurface drainage systems may discharge into open or closed drainage systems, streams, or drainage channels. Where drainage discharges into a stream or channel, a headwall or drop structure should be installed at the outlet to prevent erosion.

Wetlands

With wetland restrictions now in effect, it is suggested that state and federal regulations be consulted before planning to install a subsurface drainage system.

USE OF COMPUTERS

There is much computer software available today to accomplish many of the tasks necessary to determine runoff and to size drainageways, pipes, culverts, storage reservoirs, and so on. A computer, however, is only a tool that depends on reasonable input based on the judgment of a well prepared and knowledgeable person. Computers should only be used for computations and design if the methodology used is fully understood and the output can be evaluated for reasonableness. The designer is still responsible for potential failures and cannot blame negligence on a computer program.

SUMMARY

A number of principles of drainage system design have been presented in this chapter. However, it must be realized that storm runoff and hydraulics are highly complex subjects. The intent here is to provide a familiarity with the basic principles so that the reader may be conversant with these subjects. In most cases, the handling of storm water runoff is a collaborative effort of qualified, technically competent engineers and site designers.

EXERCISES

10.1 What is the flow velocity of a parabolic swale with a grass mixture if the cross-sectional area is 12 ft^2 and the peak rate of runoff is 36 cfs? Is this velocity acceptable for an easily erodible soil?

10.2 A drainage area consists of 8 ac of lawn and 2 ac of pavement and roofs. By the Rational method and by SCS hydrology determine the peak rate of runoff for a 10-yr storm frequency, to be used for the design of a grassed swale, if the time of concentration is 30 min.

10.3 Design a parabolic swale to handle the runoff of Exercise 10.2 if the grade is 6%, the soil is erosion-resistant, and the vegetation will be Kentucky bluegrass.

10.4 Using the Rational method, determine the peak rate of runoff for the drainage area on the site plan in Fig. 10.23 for a 10-yr storm frequency. The soil is a silt loam. Also determine the diameter of a concrete pipe (culvert) which will accommodate the runoff under the driveway.

10.5 The grading and storm drainage plan for a proposed hotel is illustrated in Fig. 10.24. Using the Rational

156 Design and Sizing of Storm Water Management Systems

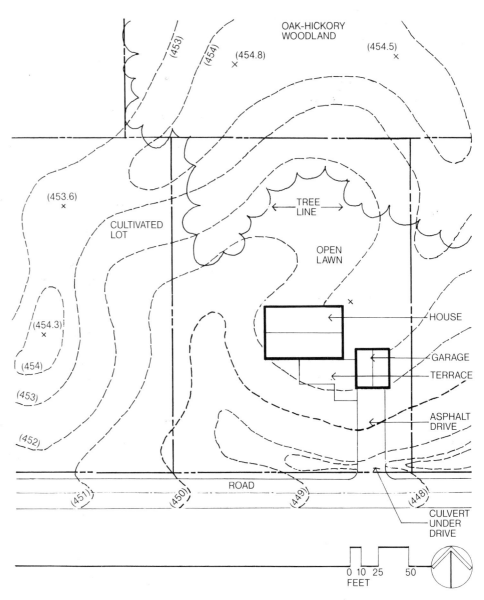

FIGURE 10.23. Site Plan for Exercise 10.4.

method, size all the pipes for the storm drainage system in terms of the following information:

a. Use a 10-yr storm frequency.

b. Each roof drain collects one-half of the roof runoff. Assume a 10-min time of concentration for this runoff.

c. Runoff coefficients are:

Roof = 0.95
Drive and parking area = 0.90
Walks = 0.80
Lawn and planting areas = 0.30

d. All pipes shall be reinforced concrete pipe (RCP) with $n = 0.013$.

e. All roads, drives, parking areas, and islands in parking areas are edged with 6-in.-high curbs.

In addition to pipe size, provide the *slope* for each pipe, the *velocity* in each pipe, and the *invert elevations* for all drainage structures. The proposed system discharges into an existing catch basin in the southeast corner with an invert elevation of 60.4 ft.

Draw a profile (horizontal scale 1 in. = 100 ft, vertical scale 1 in. = 10 ft) of the system, indicating top of frame and invert elevations, pipe size, pipe slope, and distance between structures.

10.6 (a) Examination of the topographic and land use maps of a drainage area indicates that the flow from the hydraulically most remote point to the point of concentration proceeds as follows:

300 ft of woodland with a 4% slope
200 ft of dense grass with a 3% slope
500 ft of average grass with a 2% slope
1600 ft of stream flow with 2-fps average flow velocity

Find the total time of concentration for this area, by nomograph and by SCS procedures.

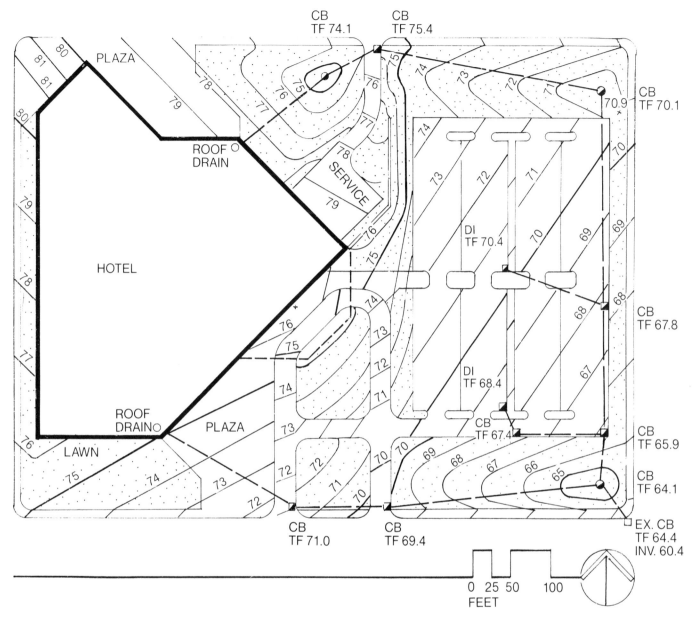

FIGURE 10.24. Site Plan for Exercise 10.5.

(b) If the area in (a) contains 125 ac, of which 35% is in woodland, 64% is in pasture, and 1% is in gravel roads, determine the peak storm runoff rate for a 25-yr frequency by the Rational method. All surface slopes of the area are less than 5%.

10.7 3.5 cfs must be conducted by a concrete circular pipe ($n = 0.013$). The gradient of the pipe must be 0.3% because of site conditions. Determine the pipe diameter required and the flow velocity in the pipe.

10.8 (a) Design a parabolic waterway for the conditions of Exercise 10.3, if the velocity cannot exceed 90% of critical velocity.

(b) Design a parabolic waterway for the conditions of Example 10.2, if the velocity cannot exceed 90% of critical velocity.

10.9 (a) Using clay tile, determine the pipe sizes required for the subsurface drainage system illustrated in Fig. 10.21, assuming all distances are doubled.

(b) Using corrugated plastic tubing, determine the pipe sizes required for the subsurface system in Fig. 10.22, assuming all distances are 1.5 times those shown.

11
Earthwork

This chapter is concerned with the sequence of earth moving and the calculation of cut and fill volumes. Methods of earth moving and types of earth-moving equipment are not discussed, since they are beyond the scope of this text. The following section defines the basic terminology associated with earthwork (Fig. 11.1).

DEFINITIONS

Finished grade: the final grade after all landscape development has been completed. It is the top surface of lawns, planting beds, pavements, and so on, and is normally designated by contours and spot elevations on a grading plan.

Subgrade: the top of the material on which the surface material such as topsoil and pavement (including base material) is placed. Subgrade is represented by the top of a fill situation and the bottom of a cut excavation. *Compacted subgrade* is a subgrade that must attain a specified density, whereas *undisturbed subgrade* is a soil that has not been excavated or changed in any way.

Base/subbase: imported material (normally coarse or fine aggregate) that is typically placed under pavements.

Finished floor elevation: the elevation of the first floor of a structure; the term may be used to designate the elevation of any floor. The relationship of the finished floor elevation to the exterior finished grade depends on the type of construction.

Cut/cutting: the process of removing soil. Proposed contours extend across existing contours in the uphill direction.

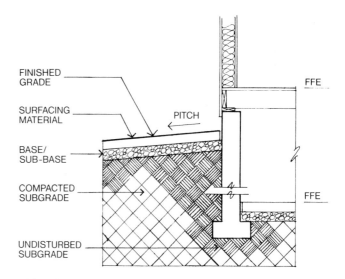

FIGURE 11.1. Grading Terminology.

Fill/filling: the process of adding soil. Proposed contours extend across existing contours in the downhill direction. When the fill material must be imported to the site, it is often referred to as *borrow*. See Fig. 11.2.

Compaction: the densification of soil under controlled conditions, particularly a specified moisture content.

Topsoil: normally the top layer of a soil profile, which may range in thickness from less than an inch to more than a foot. Because of its high organic content, it is subject to decomposition and, therefore, is not an appropriate subgrade material for structures.

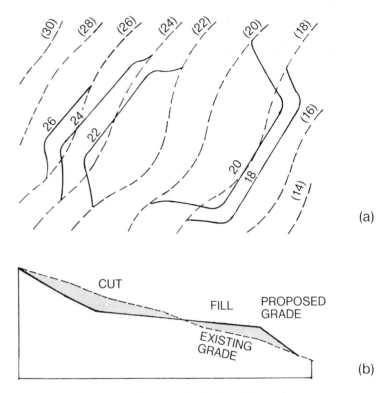

FIGURE 11.2. Cut and Fill. a. Plan indicating existing and proposed contour lines. Cutting occurs where the proposed contours move in the uphill direction; filling occurs where they move in the downhill direction.
b. Section showing where there is a change from cut to fill and where proposed grades return to existing grades. Both of these conditions are referred to as no cut–no fill.

CONSTRUCTION SEQUENCE FOR GRADING

Site Preparation

There are four areas of concern in preparing a site for grading: protection of the existing vegetation and structures that are to remain; removal and storage of topsoil; erosion and sediment control; and clearing and demolition. Of course, all four procedures may not necessarily be applicable to every project.

Protection

For the most part, protection is a phase of preparation that is self-explanatory. However, as pointed out earlier, any disturbance within the drip line of trees that are to remain should be prevented if possible. This not only refers to cutting and filling but also to the storage of materials and movement of equipment, since this will result in increased compaction and reduced aeration of the root zone of trees and shrubs.

Topsoil Removal

The site should be investigated to determine whether the quantity and quality of topsoil justify storing. The topsoil should be stripped only within the construction area and, if appropriate, stockpiled for reuse on the site. If the topsoil is to be stockpiled for a long period, it should be seeded with an annual grass to reduce loss from erosion.

Erosion and Sediment Control

Many states have enacted standards for soil erosion and sediment control, particularly for new construction. As discussed in Chapter 7, temporary control measures that divert runoff away from disturbed areas, provide surface stabilization, and filter, trap, and collect sediment should be used as appropriate. These measures should comply with all governing standards.

Clearing and Demolition

Buildings, pavements, and other structures that interfere with the proposed development must be removed before the start of construction. The same is true for interfering trees and shrubs as well as any debris that may be found on the site.

The last step in preparing a site for excavation is the placement of grade stakes. Grade stakes indicate the amount of cut or fill necessary to achieve the proposed subgrade.

Bulk Excavation

The bulk or rough grading phase is the stage at which major earth moving and shaping take place (Fig. 11.3a). The extent to which bulk excavation is necessary depends on the scale and complexity of a project. Bulk excavation includes shaping of the basic earth form and footing and foundation excavations for all structures.

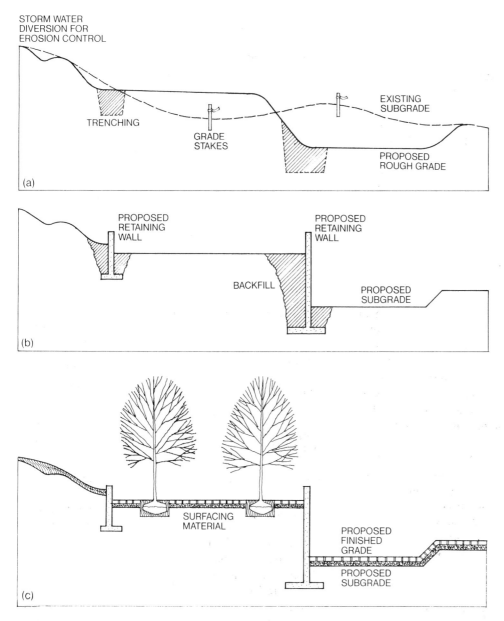

FIGURE 11.3. Grading Sequence. a. Rough grading is the phase in which major earth shaping and excavation occur. b. All utility trenches and structures are backfilled and the subgrade brought to the proper elevation during the backfilling and fine grading phases. c. Under the finished grading phase all surfacing materials, such as pavements and topsoil, are placed.

Backfilling and Fine Grading

Once rough grade has been achieved and structures have been built, finishing work may proceed. This includes backfilling excavations for structures such as retaining walls and building foundations and filling utility trenches for waterlines, sewers, and so forth. All backfills must be properly compacted to minimize future settlement problems and must be executed in a manner that does not damage utilities or structures. The last step is to make sure that the earth forms and surfaces have been properly shaped and that the subgrade has been brought to the correct elevation (Fig. 11.3b).

Surfacing

To complete the project, the surfacing material is installed. Usually the hard surfaces (i.e., pavements) are installed first and then the topsoil is placed (Fig. 11.3c). Since topsoil and pavements represent finish material, the final grades of these materials must correspond to the proposed finished grades (contours and spot elevations) indicated on the grading plan.

GRADING OPERATIONS

There are two basic ways in which a proposed grading plan may be achieved. The first is to balance the amount of cut and fill required on the site. This may be accomplished by cutting and filling in the same operation, in other words, excavating or scraping, moving, and depositing the soil in one operation. An alternative method, which may be necessary, depending on the scale and complexity of the project, is to stockpile the cut material and then place it in the fill areas as required. In either case the cut material must be suitable as fill.

The second is to import or export soil to satisfy the cut and fill requirements. This results when cut and fill do not balance on the site or when the cut material is unsuitable as fill material. Obviously, balancing cut and fill on-site is the less costly option and normally the more energy-efficient.

If it is not possible to balance cut and fill on a site, the issue arises as to which is more desirable: importing fill material (borrow) or exporting cut material. There is no general consensus among professionals, and the answer is somewhat dependent on location, scale of project, and soil conditions. However, the authors feel that it is preferable to export soil for several reasons.

First, importing soil tends to be more expensive, since it requires purchasing the material, hauling the material to the site, and then placing and properly compacting the material. Second is the condition that importing material indicates: more of the site is in fill than in cut. A fill condition is generally structurally less stable than a cut condition and is more susceptible to erosion and settlement. To reduce the potential for settlement, costly compaction methods may be necessary, or, in critical situations, special footings may have to be used for structures.

Other factors influencing the cost of grading are size and shape of site, intricacy of grading plan, and types of soil. The size, shape, and scale of a project influence labor and equipment requirements, whereas earthforms and grading tolerances influence the amount of detail and accuracy necessary in executing the design. Soil conditions also affect the type of equipment that may be used as well as the suitability of the soil for the proposed uses.

COMPUTATION OF CUT AND FILL VOLUMES

Estimates of cut and fill volumes must be made to establish construction costs and to determine whether the volumes balance or more cut or fill will be required. There are several methods for calculating volumes, three of which are discussed here: the average end area, contour area, and borrow pit (grid) methods. All the methods provide only an *approximation,* since actual cut and fill volumes are rarely the straight-edged geometric solids on which the computations are based. It should also be noted that computer programs are available for each method presented.

Average End Area Method

The average end area method is best suited for lineal construction such as roads, paths, and utility trenching. The formula states that the volume of cut (or fill) between two adjacent cross sections is the average of the two sections multiplied by the distance between them.

$$V = [(A_1 + A_2)/2] \times L \qquad (11.1)$$

where A_1 and A_2 = end sections, ft^2
V = volume, ft^3
L = distance between A_1 and A_2, ft

To apply the method, cross sections must be taken at selected or predetermined intervals. The shorter the interval between sections, the more accurate the estimate will be. Each cross section indicates the existing and proposed grades. Typically, the sections are drawn with the vertical scale exaggerated 5 to 10 times the horizontal scale. The area between the existing and proposed grades is measured, keeping cut separate from fill. Methods for measuring areas include planimeter, digitizer, geometry, and grids. The last step is to average the area of the two adjacent sections and then multiply by the distance between them to determine the volume in cubic feet. To convert to cubic yards—the standard unit of measurement for earthwork volumes—this figure must be divided by 27 ft^3/yd^3. The information is usually organized in tabular form as illustrated in Table 11.1.

Example 11.1

A portion of an existing road is to be regraded to accommodate a new culvert as illustrated in Fig. 11.4. Determine the volume of fill required. For the purposes of this problem, stations 2+00 and 3+25 represent the limits of regrading and are considered the lines of no cut or fill.

Solution. The first step is to select an interval between cross sections and to locate the sections on the plan. An interval of 50 ft is selected for this example and sections are taken at stations 2+50 and 3+00. The sections are drawn to indicate the existing and proposed grades with the vertical scale exaggerated five times. Next, the areas of the sections are measured. Usually areas are measured in square inches and then converted to square feet on the basis of the horizontal and vertical scales of the drawing. In this case, 1 in.^{2} = 30 ft × 6 ft = 180 ft^2. Since there is no cut or fill at station 2+00, the average sectional area between stations 2+00 and 2+50 is

$$\frac{0 + 101 \text{ ft}^2}{2} = 50.5 \text{ ft}^2$$

The volume is now determined by multiplying the average area by the distance between the sections:

50.5 ft^2 × 50 ft = 2525 ft^3

The volume between stations 2+50 and 3+00 is

$$\frac{101 + 40}{2} \times 50 = 3525 \text{ ft}^3$$

The last volume to be calculated is between stations 3+00 and 3+25. Again there is no change in grade at station 3+25.

Computation of Cut and Fill Volumes   163

$$\frac{40 + 0}{2} \times 25 = 500 \text{ ft}^3$$

Notice that the interval between the last two sections is only 25 ft. The data for this example are summarized in Table 11.1.

The result obtained in Example 11.1 significantly overestimates the total volume. This is true because the two end sections are actually conical or pyramidal, for which the volume is $(A/3 \times L)$, rather than $[(A_1 + A_2)/2] \times L$. A difference of 50% for the end volumes is caused by the simplification. When the conical (or pyramidal) formula is used, the volumes for Example 11.1 become 1683, 3525, and 333 ft^3, with a total of 5541 ft^3. Note that the previously computed end volumes of 2525 and 500 ft^3 are 1.5 times 1683 and 333 ft^3, respectively.

When a project has many sections, this detail is usually disregarded because of the inherent inaccuracy of volume computations. However, where only a few sections are needed (as in the example), it is advisable to use the conical formula.

Regardless of the formulas used, the average end area method tends to overestimate volumes.

Contour Area Method

The contour area method is appropriate for large, relatively uncomplicated grading plans and may also be used to calculate volumes of water in ponds and lakes. To apply this method, the first step is to establish the line of no cut or fill and then to separate the area of cut from the area of fill. The next step is to measure the horizontal area of change for *each* contour line within the no cut–no fill limit, keeping areas of cut separate from areas of fill. In other words, measure the area bounded by the same numbered existing and proposed contour lines. Finally, the volumes of cut and fill can be calculated by applying the following formula:

$$V = \frac{A_1 h}{3} + \frac{(A_1 + A_2)h}{2} + \ldots \frac{(A_{n-1} + A_n)h}{2} + \frac{A_n h}{3} \quad (11.2)$$

where $A_1, A_2 \ldots A_n$ = area of horizontal change for each contour, ft^2

h = vertical distance between areas, ft

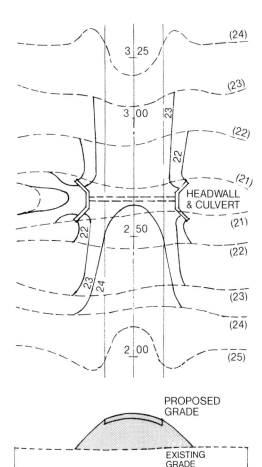

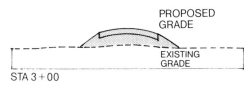

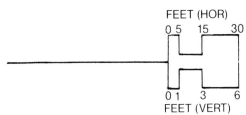

FIGURE 11.4. Plan and Sections for Example 11.1.

TABLE 11.1. Data for Example 11.1

Station	Area (ft²)	Average (ft²)	Distance (ft)	Volume (ft³)
2+00	0			
		50.5	50	2525
2+50	101			
		70.5	50	3525
3+00	40			
		20.0	25	500
3+25	0			
				6550

164   Earthwork

The first and last terms on the right-hand side of the equation are considered as conical or pyramidal solids, as shown by the shaded areas in section in Fig. 11.5. The volume of such a solid $(A/3)h$ was given in the preceding example. If the altitude h is equal to the contour interval i, the equation can be simplified as follows:

$$V = i(5/6A_1 + A_2 + A_3 + \ldots 5/6A_n) \tag{11.3}$$

Because of the approximate nature of computing earthwork, the equation can be further simplified to

$$V = i(A_1 + A_2 + A_3 + \ldots A_n) \tag{11.4}$$

However, using this final form of the equation will result in overestimated earthwork volumes. For ease of computation the information should be organized as shown in Table 11.2.

Example 11.2

Figure 11.5 illustrates a slope which has been regraded to accommodate a small plateau area. Using the contour area method, calculate the volumes of cut and fill required.

Solution. As noted, the first step is to delineate the extent of the earthwork by a no cut–no fill line and to sepa-

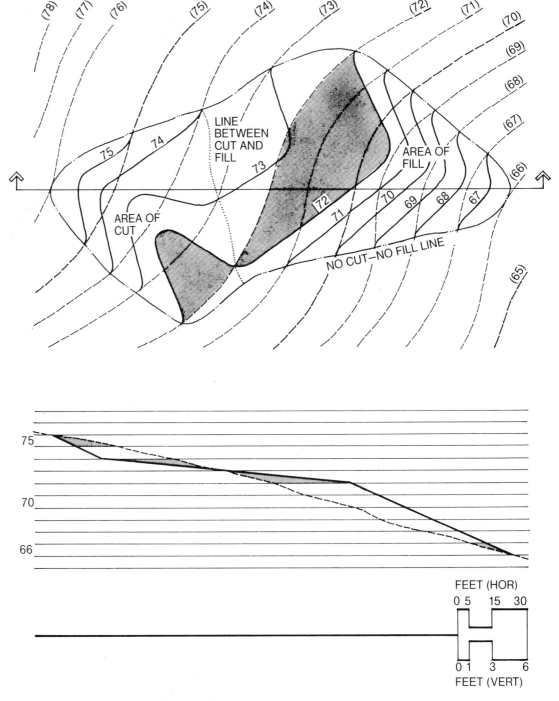

FIGURE 11.5. Plan and Section for Example 11.2.

rate the area of cut from the area of fill; then the area of change is measured for each contour line. This is demonstrated by the shaded areas for the 72-ft contour line. The areas for each contour line are recorded in Table 11.2. The last step is to multiply the total areas by the contour interval, in this case, 1 ft, and divided by 27 ft³/yd³ to convert the volumes to cubic yards.

$$\frac{1282 \times 1}{27} = 47.48 \text{ yd}^3 \text{ cut}$$

$$\frac{3140 \times 1}{27} = 116.30 \text{ yd}^3 \text{ fill}$$

TABLE 11.2. Contour Area Measurements

Contour number	Area of cut (ft²)	Area of fill (ft²)
67	0	60
68	0	175
69	0	324
70	0	451
71	0	780
72	308	990
73	388	360
74	410	0
75	176	0
Total	1282	3140

Borrow Pit Method

The borrow pit method, sometimes referred to as the grid method, is appropriate for complex grading projects and urban conditions. Existing elevations are determined at each grid intersection on the site by "borrow pit leveling," normally done in preparation for contour mapping, as described in Chapter 2. If such elevations are not available but a contour map of the project site has been prepared, a grid is placed over the area to be regraded. Care must be taken in determining the size and location of the grid on the site. The estimate becomes more accurate as the size of the grid decreases, and in some cases it may even be appropriate to break up the area into two or more parts, each with a different size grid. Existing and proposed grades are determined at each grid intersection by interpolation and the difference between elevations calculated. A notational system should be used to distinguish fill from cut, such as F and C. From this point there are two ways to proceed.

The first approach is to apply the borrow pit method on a cell by cell basis. An average change in elevation is calculated for each cell by determining the difference in elevation for all four corners of each cell, as illustrated in Fig. 11.6, adding all four differences and dividing by 4. The volume is calculated by adding all the averaged values together, keeping cut and fill separate, and multiplying by the area of one grid cell.

Example 11.3

The grid of spot elevations illustrated in Fig. 2.5 is shown in Fig. 11.7 with the resulting contour lines. For this problem a rectangular area bounded by corners B2, B5, D2, and D5 is to be excavated with vertical sides to a finished elevation of 95.0. Calculate the volume of excavation using the borrow pit method. For identification purposes the six cells are numbered in Fig. 11.7.

Solution. The first step is to determine the difference in elevation between the existing and proposed grades. This is demonstrated for cell no. 1 as follows (see Fig. 11.6a):

Corner: B2 98.5 − 95.0 = 3.5 ft

B3 99.4 − 95.0 = 4.4 ft

C2 97.6 − 95.0 = 2.6 ft

C3 98.3 − 95.0 = 3.3 ft

Next the differences are totaled and divided by 4 to calculate the average.

$$\frac{3.5 + 4.4 + 2.6 + 3.3}{4} = \frac{13.8}{4} = 3.45 \text{ ft avg}$$

The same procedure is applied to the remaining five cells with the following results:

Cell no. 2: 3.475 ft avg
Cell no. 3: 2.50 ft avg
Cell no. 4: 2.30 ft avg
Cell no. 5: 2.35 ft avg
Cell no. 6: 1.775 ft avg

The final step is to add up the averages for all the cells and to multiply by the area of one cell (100 ft × 100 ft = 10,000 ft²).

$$3.45 + 3.475 + 2.50 + 2.30 + 2.35 + 1.775 = 15.85 \text{ ft}$$

$$15.85 \times 10,000 = 158,500 \text{ ft}^3$$

$$\frac{158,500}{27} = 5870.37 \text{ yd}^3$$

Since earthwork computations are only approximate, this may be rounded to 5900 yd³.

The second approach is derived by simplifying the equation of the first approach by common factoring. The advantage of this procedure is that it reduces the number of calculations required.

$$V = \frac{A}{4} \times (1h_1 + 2h_2 + 3h_3 + 4h_4) \qquad (11.5)$$

where: V = volume of cut (or fill), ft³
 A = area of one grid cell, ft²
 h_1 = sum of the cuts (or fills) for all grid corners common to one grid cell
 h_2 = sum of the cuts (or fills) for all grid corners common to two grid cells
 h_3 = sum of the cuts (or fills) for all grid corners common to three grid cells
 h_4 = sum of the cuts (or fills) for all grid corners common to four grid cells

166   Earthwork

(a)   (b)

FIGURE 11.6. a. Example of single grid cell for borrow pit method. b. Borrow pit leveling on a project site. Plans for a borrow pit grid are illustrated in Figs. 2.5 and 11.7.

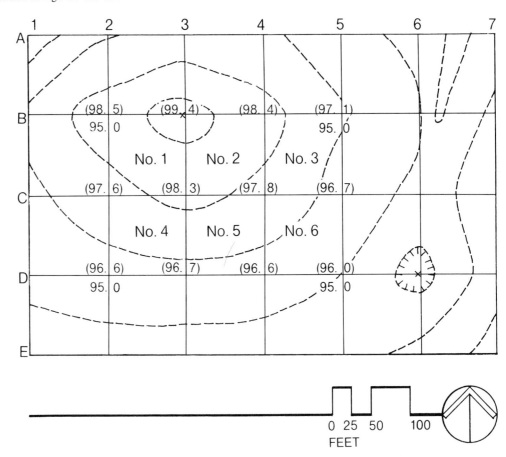

FIGURE 11.7. Plan for Example 11.3.

This process is demonstrated in the following example.

Example 11.4

Using Example 11.3, the volume of excavation is now calculated by the simplified equation.

Solution. There are four corners that appear in only *one* grid cell of the area to be excavated: B2, B5, D2, and D5. The depth of cut for these corners is obtained by subtracting the proposed finished elevation from the existing elevation: $98.5 - 95.0 = 3.5$ ft for B2. The remaining three cuts are 2.1, 1.6, and 1.0. The sum of h_1 is $3.5 + 2.1 + 1.6 + 1.0$, which equals 8.2 ft.

Proceeding clockwise, the corners common to two grid cells of the excavated area are B3, B4, C5, D4, D3, and C2. The depths of cut for these corners are 4.4, 3.4, 1.7, 1.6, 1.7, and 2.6, again obtained by subtracting the proposed finished elevation, 95.0 ft, from the existing grades. The sum of h_2 is 15.4 ft.

There are two corners common to four grid cells within the area to be graded: C3 and C4. The sum of h_4 is 3.3 + 2.8, which equals 6.1 ft. There are no corners common to three grid cells. Since all 12 corners of the grid have been accounted for, the values may be substituted into the formula. Again each cell measures 100 ft by 100 ft; therefore, the area of each cell equals 10,000 ft².

$$V = \frac{A}{4}(1h_1 + 2h_2 + 3h_3 + 4h_4)$$

$$= \frac{10,000}{4}(1 \times 8.2 + 2 \times 15.4 + 3 \times 0.0 + 4 \times 6.1)$$

$$= 2500\,(8.2 + 30.8 + 0.0 + 24.4)$$

$$= 158,500 \text{ ft}^3$$

$$= \frac{158,500}{27} = 5870.37 \text{ yd}^3$$

For estimating purposes this may again be rounded upward to 5900 yd³.

The method works the same way for fill, except that the height of fill required for each corner is determined by subtracting the existing elevation from the proposed finished elevation. Where a project consists partially of cut and partially of fill, the cut and fill are determined separately and the net volume of cut or fill can then be computed.

Adjustment of Cut and Fill Volumes

Two adjustments must be made in determining cut and fill volumes. The first is concerned with surface materials; the second involves compaction and shrinkage of soil volumes.

For estimating purposes, cut and fill volumes are determined between existing and proposed subgrades, *not* between existing and proposed finished grades. However, contour lines and spot elevations on grading plans and topographic surveys usually indicate finished grade conditions. As a result, compensation must be made for both the existing surface material to be removed and the proposed surfacing material to be installed. This may be accomplished in a variety of ways, a few of which will be discussed here. It is important, however, to understand the following basic principles.

1. In cut, proposed surfacing material (including pavement and topsoil) *increases* the amount of excavation required.
2. In fill, proposed surfacing material *decreases* the amount of borrow required.
3. In cut, the removal of existing pavement or stripping of topsoil *decreases* the volume of soil to be removed.
4. In fill, the removal of existing pavement or stripping of topsoil *increases* the volume of soil to be placed.

These principles are illustrated in Fig. 11.8. It should be noted that where the depths of the proposed and existing surface materials are the same, they are self-compensating and no adjustment is required. With both the average end area and borrow pit methods it is possible to incorporate the adjustment directly into the volume calculations. For the average end area method the cross sections may be drawn to indicate the existing and proposed subgrades rather than finished grades. For the borrow pit method, the spot elevations indicated at the grid corners could be based on the proposed and existing subgrades.

Another alternative is simply to measure the plan area of the surface material, keeping areas within cut separate from areas within fill, and multiply by the depth of the material to calculate the volume. The volume may be added or subtracted to the cut or fill volumes as described in the principles given. This technique is illustrated in Example 11.5.

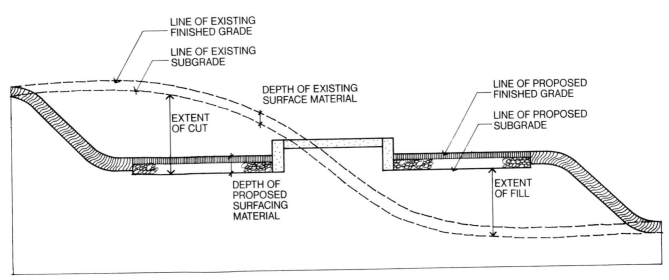

FIGURE 11.8. Relationship of Existing and Proposed Surfacing Materials to Cut and Fill Volumes. Where existing and proposed finished grade elevations, rather than subgrade elevations, are used to compute volumes, make the following adjustments: Gross cut volume − existing surfacing material volume + proposed surfacing material volume = adjusted cut volume. Gross fill volume + existing surfacing volume − proposed surfacing material volume = adjusted fill volume.

Example 11.5

Again, using the data from Example 11.3, the volume of excavation is to be adjusted, on the basis of the following information. An existing 6-in. layer of topsoil must be stripped before excavation can begin. The proposed 95.0-ft finished elevation is the top of a 6-in. concrete slab with a 4-in. gravel base.

Solution. The topsoil to be stripped reduces the gross volume of excavation. The volume of topsoil is the area (60,000 ft²) multiplied by the depth (0.5 ft).

$60,000 \times 0.5 = 30,000$ ft³

$\dfrac{30,000}{27} = 1111.11$ yd³, or 1110 yd³

On the other hand, the depth of the excavation must be increased by 10 in. (0.83 ft) to accommodate the concrete slab and base course. The resulting increased cut volume is

$60,000 \times 0.83 = 49,800$ ft³

$\dfrac{49,800}{27} = 1844.44$ yd³ or 1840 yd³

The final adjusted volume is determined as follows:

5900 yd³ (gross cut) + 1840 yd³ (proposed surfacing) − 1110 yd³ (existing topsoil) = 6630 yd³ adjusted volume

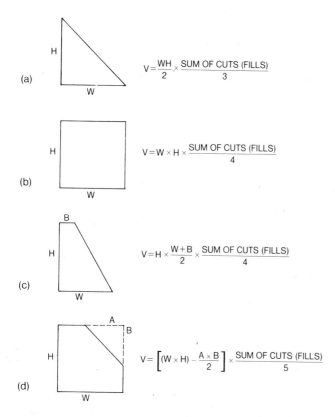

FIGURE 11.9. Borrow Pit Volume Formulas. a. Triangular areas. b. Square and rectangular areas. c. Trapezoidal areas. d. Pentagonal areas.

The second factor affecting cut and fill volumes is the change in soil volume as a result of compaction and shrinkage. Usually *in-place* volumes of cut, referred to as bank yards, yield less than their volume in fill by 10% to 20%, depending on soil type and compaction techniques. This means that 100 yd³ of cut yields approximately only 80 to 90 yd³ of fill. Therefore, to balance cut and fill on a site, a 100-yd³ fill would require approximately 110 to 125 yd³ of cut. For cost estimating purposes, fill volumes, referred to as compacted yards, should be increased by 10% to 20%, depending on the soil, to determine the actual quantity of borrow required.

Balancing of Cut and Fill Volumes

On some projects it may be desirable, or even required, that all grading be self-contained on the site, that is, that no soil can be imported to or exported from the site. On such projects, the line of no cut–no fill on the grading plan will result in areas with shapes other than square or rectangular. The borrow pit method described previously can be applied only when all grid cells are identical squares or rectangles. Where the areas of the cells are not equal, the volume for each area must be determined separately.

The basic approximation for computing cut or fill is that the volume is the product of the mapped (horizontal) area multiplied by the average of the cuts (or fills) at each corner of that area. Some of these cuts (or fills) may be zero, but they *must* be included in averaging. Figure 11.9 illustrates the variety of geometrical areas that may occur on a grading plan and formulas for the corresponding volumes.

Example 11.6

Figure 11.10 represents a project that will be partially in fill and partially in cut. The number at each corner indicates the height of fill or depth of cut required. Determine the volume of fill and cut for each area and the total volumes of fill and cut. All distances are as shown on the plan.

Solution. Volumes 1, 2, 3, and 4 are in cut and 5, 6, and 7 are in fill. The volumes for each area will be computed separately according to the formulas given in Fig. 11.9.

Volume no. 1 $= W \times H \times \dfrac{\text{sum of cuts}}{4}$

$= 100 \times 100 \times \dfrac{4.8 + 3.5 + 0.8 + 4.2}{4}$

$= \dfrac{10,000 \times 13.3}{4} = 33,250$ ft³

$= \dfrac{33,250 \text{ ft}^3}{27 \text{ ft}^3/\text{yd}^3} = 1231.48$ yd³

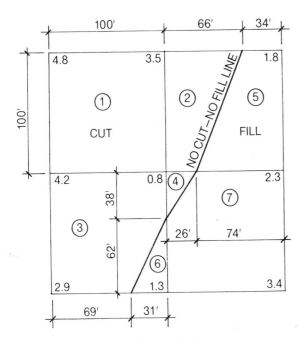

FIGURE 11.10. Plan for Example 11.6.

Volume no. 2 = $H \times \dfrac{(W + B)}{2} \times \dfrac{\text{sum of cuts}}{4}$

$= 100 \times \dfrac{(66 + 26)}{2} \times \dfrac{3.5 + 0 + 0 + 0.8}{4}$

$= \dfrac{4600 \times 4.3}{4} = 4945 \text{ ft}^3 = 183.15 \text{ yd}^3$

Note that the zeros used in the sum of the cuts represent the corners at the no cut–no fill line.

Volume no. 3 = $[(W \times H) - (A \times B)/2] \times \dfrac{\text{sum of cuts}}{5}$

$= [(100 \times 100) - (62 \times 31)/2] \times \dfrac{4.2 + 0.8 + 0 + 0 + 2.9}{5}$

$= \dfrac{9039 \times 7.9}{5} = 14281.62 \text{ ft}^3 = 528.95 \text{ yd}^3$

Volume no. 4 = $\dfrac{(W \times H)}{2} \times \dfrac{\text{sum of cuts}}{3}$

$= \dfrac{(26 \times 38)}{2} \times \dfrac{0.8 + 0 + 0}{3}$

$= \dfrac{494 \times 0.8}{3} = 131.73 \text{ ft}^3 = 4.88 \text{ yd}^3$

Similarly, the volumes of fill are computed as follows:

Volume no. 5 = $\dfrac{5400 \times 4.1}{4} = 5535 \text{ ft}^3 = 205.00 \text{ yd}^3$

Volume no. 6 = $\dfrac{961 \times 1.3}{3} = 416.43 \text{ ft}^3 = 15.42 \text{ yd}^3$

Volume no. 7 = $\dfrac{9506 \times 7.0}{5} = 13308.4 \text{ ft}^3 = 492.90 \text{ yd}^3$

The total volume of the cut is 1231.48 + 183.15 + 528.95 + 4.88 = 1948.46 yd³, which may be rounded to 1950 yd³. The volume of fill is 205.00 + 15.42 + 492.90 = 713.32 yd³, which is approximately 710 yd³.

It may be noted that the volume of cut is 1950/710, or more than 2.7 times the volume of fill, which may also be stated as a cut/fill ratio of 2.7. Even considering shrinkage (settling and compaction), it is evident that there will be a considerable amount of extra soil. A ratio of 1.1 to 1.2 ft³ of cut to each 1 ft³ of fill required is often used where a balance of volumes is desired. The next example will demonstrate how this may be accomplished.

Example 11.7

Assuming that the desired final grade of the project in Fig. 11.10 is to be level, adjust all cuts and fill heights to achieve a cut/fill ratio of 1.2.

Solution. Achieving a 1.2 cut/fill ratio basically requires a trial and error procedure. Since there is an excess of cut (or since the proposed grade is too low), the final grade must be raised, resulting in less cut and more fill. As a result, the *area* of cut will decrease while the *area* of fill will increase. Also, for every 0.1 ft that the final grade is raised, 0.1 ft must be subtracted from each cut depth and added to each fill height.

For a preliminary estimate of volumes, it can be assumed that the cut (or fill) occurring on an outside corner is the average cut (or fill) for one-quarter of the area of a grid square. The cuts (or fills) common to two grid squares may be assumed to be the average cuts (or fills) for two-quarters, or one-half, of the area of a grid square; the cuts (or fills) common to three squares account for three-quarters of a grid square; and cuts (or fills) common to four grid squares may be averaged over four-quarters, or the area or one full grid square. This is illustrated in Fig. 11.11. These assumptions make it possible to make preliminary computations with new trial cut depths or fill heights without actually computing volumes, which would be more laborious. Also, as a result of the change in cut and fill *areas,* the actual location of the final no cut–no fill line is not yet known. In the preliminary computations, each cut (or fill) on an outside corner will have one-quarter weight, each one common to two grid squares will have one-half weight, each one common to three grid squares will have three-quarters weight, and each one common to four grid squares will have one full weight.

For a trial grade the weighted cuts and fills are determined. The sum of the weighted cuts is compared to the sum of the weighted fills. When the ratio of these sums is close to the desired cut/fill ratio, actual volumes are computed.

In this example, since there is a large excess of cut, the first trial will raise the proposed grade 0.6 ft, so that each cut will be 0.6 ft less and each fill 0.6 ft more, as shown in Fig. 11.12. The sum of the weighted cuts is 4.2/4 +

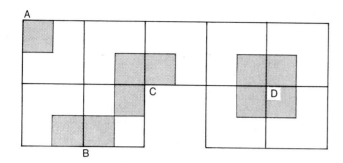

FIGURE 11.11. Approximate Average Cut and Fill Values. Rather than average the corner values for each grid cell, the cut or fill value at each grid intersection may be assumed to be the average value for the adjacent grid cell area. Thus point A is the average for one-quarter of a cell, B is the average for one-half of a cell, C is the average for three-quarters of a cell, and D is the average for one full cell.

$2.9/2 + 3.6/2 + 0.2 + 2.3/4 = 5.075$ ft. The sum of the weighted fills is $2.4/4 + 2.9/2 + 1.9/2 + 4.0/4 = 4$. Therefore, the trial cut/fill ratio is $5.075/4 = 1.27$. This is close enough to the desired 1.2 ratio to compute the actual volumes.

To do this, the first step is to locate the no cut–no fill line wherever there is a change from cut to fill. The point at which the no cut–no fill line crosses a grid line can be determined by the following proportion:

$$\frac{d_1}{L} = \frac{C}{C + F} \quad (11.6)$$

Also

$$d_2 = L - d_1 \quad (11.7)$$

where: d_1 = distance to the point of no cut–no fill from the grid corner in cut
L = distance between grid corners (100 ft, in this example)
C = depth of cut
F = fill height
d_2 = distance to the point of no cut–no fill from the grid corner in fill

The relationship can also be expressed as follows, if it is more convenient to proceed from the grid corner that is in fill:

$$\frac{d_2}{L} = \frac{F}{F + C} \quad (11.8)$$

And

$$d_1 = L - d_2 \quad (11.9)$$

Using these relationships, the distance for the lower left is computed as follows:

$$\frac{d_1}{100} = \frac{2.3}{2.3 + 1.9}$$

$$d_1 = \frac{100 \times 2.3}{4.2} = 54.76 \text{ or } 55 \text{ ft}$$

$$d_2 = 100 - 55 = 45 \text{ ft}$$

Similarly, the other distances are determined as follows:

$$d_2 = 100 \times \frac{1.9}{1.9 + 0.2} = 90 \text{ ft } (d_1 = 10 \text{ ft})$$

$$d_1 = 100 \times \frac{0.2}{0.2 + 2.9} = 6 \text{ ft } (d_2 = 94 \text{ ft})$$

$$d_1 = 100 \times \frac{2.9}{2.9 + 2.4} = 55 \text{ ft } (d_2 = 45 \text{ ft})$$

The location of the no cut–no fill line is shown in Fig. 11.13. The line has shifted to the left, and, as noted previously, the area of cut has been reduced and the area of fill increased. The volumes of cut and fill are computed as before with the following results:

Cut

Volume no. 1 = $100 \times 100 \times \frac{10.9}{4} = 27{,}250$ ft³
$= 1009.26$ yd³

Volume no. 2 = $100 \times \frac{55 + 6}{2} \times \frac{3.1}{4} = 2363.75$ ft³
$= 87.55$ yd³

Volume no. 3 = $[(100 \times 100) - \frac{90 \times 45}{2}] \times \frac{6.1}{5}$
$= 9729.5$ ft³
$= 360.35$ yd³

Volume no. 4 = $\frac{6 \times 10}{2} \times \frac{0.2}{3} = 2.0$ ft³ $= 0.07$ yd³

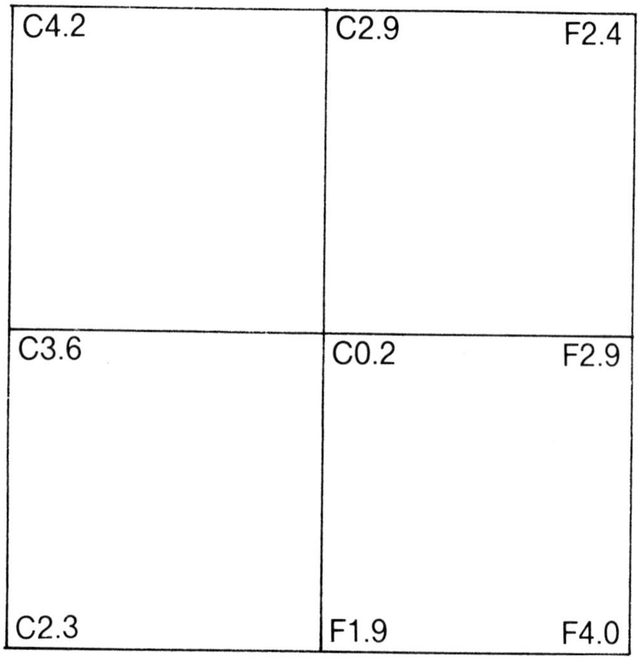

C4.2	C2.9	F2.4
C3.6	C0.2	F2.9
C2.3	F1.9	F4.0

FIGURE 11.12. Plan.

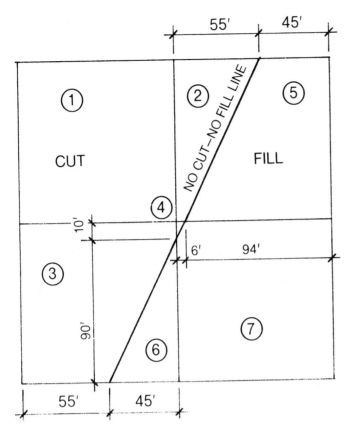

FIGURE 11.13. No Cut–No Fill Line.

Fill

Volume no. 5 = $6950 \times \dfrac{5.3}{4}$ = 9208.75 ft³ = 341.06 yd³

Volume no. 6 = $2025 \times \dfrac{1.9}{3}$ = 1282.5 ft³ = 47.5 yd³

Volume no. 7 = $9970 \times \dfrac{8.8}{5}$ = 17,547.2 ft³ = 649.9 yd³

$\dfrac{\text{Total volume of cut}}{\text{Total volume of fill}} = \dfrac{1457.23}{1038.46} = 1.4$

This is more than the preliminary ratio of 1.27 and is still too great. The final grade will be raised again by 0.1 ft as shown in Fig. 11.14. The resulting volumes are as follows:

Cut

Volume no. 1 = 972.22 yd³

Volume no. 2 = 75.19

Volume no. 3 = 331.67

Volume no. 4 = 0.01

  Total cut = 1379.09 yd³

Fill

Volume no. 5 = 366.67 yd³

Volume no. 6 = 56.30

Volume no. 7 = 673.57

  Total fill = 1096.54 yd³

The cut/fill ratio is now 1.26, which is still a little high. However, raising the final grade another 0.1 ft produces a cut/fill ratio of 1.06. Therefore, the 1.26 ratio is accepted, since it is very difficult to grade to tolerances of less than 0.1 ft.

Although the desired final surface was to be level in Example 11.7, the method used can be applied to any required final plane surface, level or sloping. The mapped (horizontal) distances should be used to determine the areas for volume computations and to locate the no cut–no fill lines.

Another point that aids in achieving a balance is to understand the relationship of volume to depth and area. Raising or lowering large areas only a few inches may significantly change cut and fill volumes. For instance, if the grade over a 1-ac area is raised or lowered by 4 in., the change in volume is greater than 500 yd³. Establishing simple area-to-volume relationships for different depths makes the task of balancing cut and fill easier.

EXERCISES

11.1 Using the average end area method, calculate the volumes of cut and fill for the cross sections (horizontal and vertical scales as shown) indicated in Fig. 11.15. Adjust the volumes on the basis of the following information:

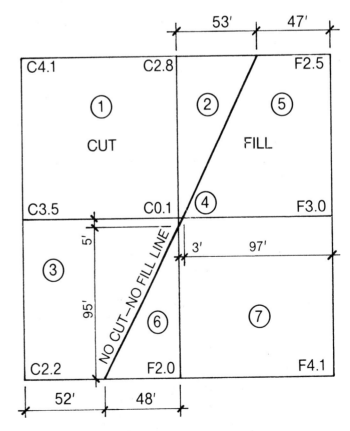

FIGURE 11.14. Final Plan.

172   Earthwork

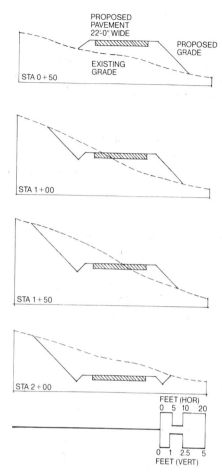

FIGURE 11.15. Sections for Exercise 11.1.

a. Depth of existing topsoil: 6 in.
b. Depth of proposed topsoil: 6 in.
c. Depth of proposed pavement: 6 in.

Stations 0+00 and 2+35 are points of no cut and no fill. Also indicate the approximate volume of *excess topsoil*.

11.2 Using the contour area method, determine the *adjusted* volumes of cut and fill for the plan in Fig. 11.16 in terms of the following information:

a. Depth of existing topsoil: 4 in.
b. Depth of topsoil replaced in disturbed areas: 8 in.
c. Depth of proposed pavement: 6 in.

How much soil must be imported or exported from the site? Also determine the volume of topsoil to be replaced.

11.3 Using the grading plan in Fig. 5.26, construct a grid of 20-ft by 20-ft cells starting at the northeast corner over the entire site. Determine the *existing and proposed subgrades* at each of the grid intersections based on the following:

a. Depth of existing topsoil: 4 in.
b. Depth of proposed topsoil: 6 in.
c. Depth of proposed pavement: 8 in.
d. Subgrade of building 1.0 ft below FFE

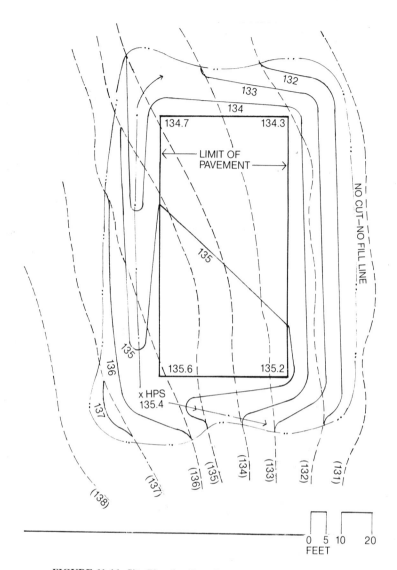

FIGURE 11.16. Site Plan for Exercise 11.2.

Finally, using the borrow pit method, compute the volumes of cut and fill for this site.

11.4 The area illustrated in Fig. 11.17 is to be leveled without importing or exporting soil. The desired cut/fill ratio is 1.2.

a. Determine the final elevation
b. Determine the volume of cut
c. Determine the volume of fill
d. Determine the final cut-to-fill ratio
e. Show the location of the no cut–no fill line

11.5 The existing elevations at the grid corners of a 100-ft by 200-ft area are given (Fig. 11.18). The site plan for the project requires that line *B–B* be level and that there be a 2% slope downward from *B–B* toward *A–A* and toward *C–C*. Also, the cut/fill ratio is to be as close as possible to 1.2. No soil is to be brought to or taken from the site. Draw the final grading plan, showing the elevation at each grid corner. Show the no cut–no fill line with all distances indicated. Determine the volumes of cut and fill and the final cut/fill ratio.

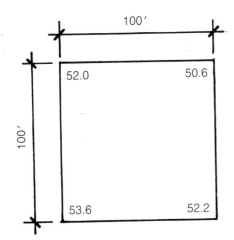

FIGURE 11.17. Plan for Exercise 11.4.

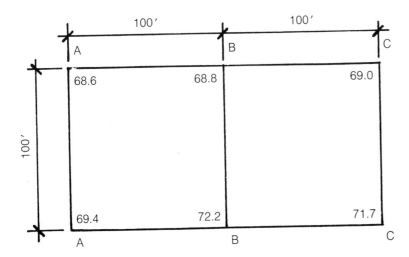

FIGURE 11.18. Plan for Exercise 11.5.

12
Horizontal Road Alignment

Generally, landscape architects and site designers are involved with the design of low-speed residential and park roads, entry and service drives, and parking areas. Involvement with high-speed roads is limited to corridor or route selection but, for the most part, is not concerned with highway engineering. The purpose of the next two chapters is to present the basic engineering necessary to lay out roads and drives in the landscape. Visual, experiential, and environmental issues; traffic engineering; and traffic management techniques are not addressed, since they are beyond the scope of this text.

To create safe, enjoyable, and easily maneuverable vehicular circulation, roads must be engineered in both the horizontal and vertical planes. The horizontal plane is concerned with the alignment of roads *through* the landscape, referred to as *horizontal alignment*. The vertical plane is concerned with the alignment of roads *over* the landscape, which is accommodated by vertical curves (for further discussion, see Chapter 13).

TYPES OF HORIZONTAL CURVES

In the horizontal plane, road alignment consists of two basic geometric components: the straight line, or *tangent*, and the *curve*. The tangent is the most common element of road alignment. It represents the shortest distance between two points, and laying it out is easy. In flat or featureless terrain or in urban grid situations its use may be appropriate. However, because of its predictability and limiting viewpoint, a straight line in the landscape may become aesthetically and experientially uninteresting. To create interest and respond to natural features such as topography and vegetation, tangents may change direction. Two basic types of curve are used to accommodate this change in direction.

Circular Curves

As the name implies, circular curves are circular arcs with a constant radius. They are easy to calculate and lay out and may be used in four basic configurations (Fig. 12.1).

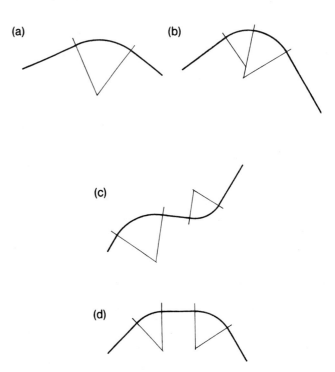

FIGURE 12.1. **Types of Horizontal Curves.** a. Simple curve with a single radius. b. Compound curve. c. Reverse curve. d. Broken-back curve.

Simple curve: a curve with a single radius, which is the most common configuration for low-speed roads.

Compound curve: a curve consisting of two or more radii in the same direction. For continuity and ease of handling, the difference in length of the radii should not be greater than 50%.

Reverse curve: two arcs in *opposite* directions. Usually a tangent is required between the two arcs, the length of which is determined by the design speed of the road.

Broken-back curve: two curves in the *same* direction connected by a tangent. Where the tangent distance is relatively short, these curves may be uncomfortable to maneuver and visually disjointing. Therefore, if possible, this condition should be prevented by using one larger curve.

Spiral Transitional Curves

The normal path through a curve at high speeds is not circular, but through a series of curves with a constantly changing radius. This phenomenon is reflected in road alignment design by the use of spiral curves. The major disadvantage of this type of curve is that it is more difficult to calculate and lay out. For the design speeds and scale of roads in which landscape architects are involved, circular curves are sufficient. Therefore, spiral curves will not be discussed here.

CIRCULAR CURVE ELEMENTS

Figure 12.2 illustrates the elements of circular curves, which are defined as follows:

Point of curvature (PC): the point that marks the beginning of the curve at which the road alignment diverges from the tangent line in the direction of stationing.

Point of tangency (PT): the point that marks the end of the curve at which the road alignment returns to a tangent line in the direction of stationing.

Point of intersection (PI): the point at which the two tangent lines intersect.

Included angle (I): the central angle of the curve, which is *equal to the deflection angle between the tangents*.

Tangent distance (T): the distance from the PI to either the PC or the PT. These distances are *always* equal for simple circular curves.

Radius (R): the radius of the curve.

Length of curve (L): the length of the arc from PC to PT.

Chord (C): the distance from PC to PT measured along a straight line.

Center of curve (O): the point about which the included angle *I* is turned.

CIRCULAR CURVE FORMULAS

Using basic trigonometric and geometric relationships in conjunction with the preceding definitions, the following formulas may be derived for circular curves:

$$T = R \tan \frac{I}{2} \qquad (12.1)$$

$$C = 2R \sin \frac{I}{2} \qquad (12.2)$$

Note that the radius is always perpendicular to the tangent line at PC and PT. The two relationships can be derived from Fig. 12.2. The arc length L can be computed as part of the total circumference of a circle by proportion as follows:

$$\frac{L \text{ (length of curve)}}{2\pi R \text{ (circumference of circle)}} = \frac{I \text{ (included angle)}}{360° \text{ (total degrees in circle)}}$$

$$L = \frac{(2\pi R) \times I}{360} \qquad (12.3)$$

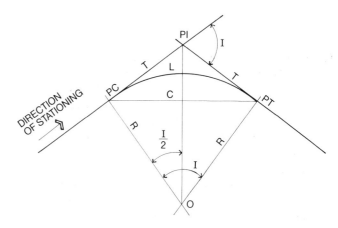

FIGURE 12.2. Elements of Horizontal Curves.

Example 12.1

Two tangent lines intersect at point *B* as illustrated in Fig. 12.3. The bearing of tangent line *AB* is N76°30′E, while the bearing of tangent line *BC* is S15°20′E. A circular curve with a 100-ft radius is to be constructed to connect the two tangent lines. Find the included angle, the tangent distance (the distance from point *B* to the beginning or the end of the curve), the length of the curve (arc), and the chord distance.

Solution. With the aid of Fig. 12.4, the deflection angle between the tangent lines is determined to be 88°10′. As previously noted, the deflection angle and the included angle are equal; therefore, the included angle for this circular curve is 88°10′.

The formula for calculating the tangent distance is

$$T = R \tan \frac{I}{2}$$

By substituting the known values the equation becomes

$$T = 100 \tan \frac{88°10'}{2}$$

Circular Curve Formulas 177

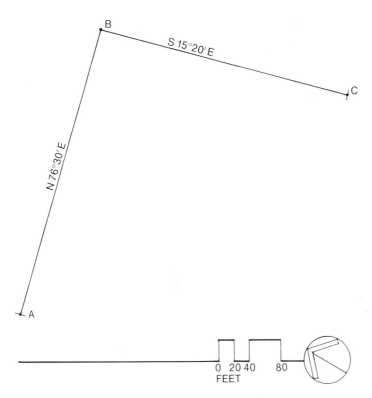

FIGURE 12.3. Tangent Line Bearings for Example 12.1.

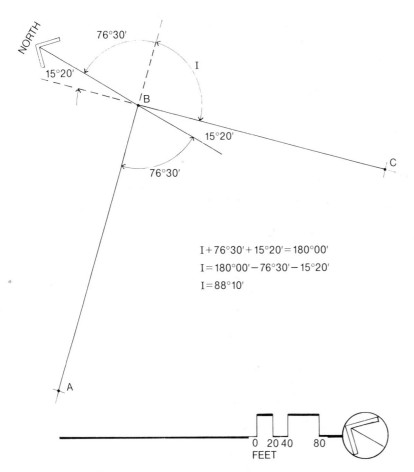

$$I + 76°30' + 15°20' = 180°00'$$
$$I = 180°00' - 76°30' - 15°20'$$
$$I = 88°10'$$

FIGURE 12.4. Diagram for Calculating the Deflection and Included Angles.

$= 100 \tan 44°5'$
$= 100 \times 0.9685$
$= 96.85$ ft

The length of the curve is determined by substituting the known radius into the following proportion:

$$\frac{L}{2\pi R} = \frac{I}{360°}$$

$$\frac{L}{2\pi 100} = \frac{88°10'}{360°}$$

$$L = 2\pi 100 \times \frac{88.167°}{360}$$

$= 153.88$ ft

Note that the minutes (and seconds if applicable) of the included angle are converted to decimals of degrees before using them in multiplication or division.

Finally the chord length is determined by the following equation:

$$C = 2R \sin \frac{I}{2}$$
$= 2 \times 100 \sin 44°5'$
$= 2 \times 100 \times 0.6957$
$= 139.14$ ft

DEGREE OF CURVE

Some organizations such as highway departments simplify their computations for curve layouts and establish minimum curve standards by using the designation *degree of curve*. There are two definitions for degree of curve. The *chord definition* defines it as the angle subtending a 100-ft chord, the *arc definition* defines it as the angle subtending a 100-ft arc. This discussion will be limited to the latter definition, since it is the one more commonly used today.

From the relationship illustrated in Fig. 12.5, the following proportion may be established (the same proportion as that used to determine length of curve):

$$\frac{100}{D} = \frac{2\pi R}{360°} \quad (12.4)$$

where: D = subtended angle (degree of curve)
R = radius of curve

Thus

$$R = \frac{100 \times 360°}{2\pi D}$$

$$R = \frac{5729.578}{D} \approx \frac{5730}{D}$$

Therefore, degree of curve D is inversely proportional to the radius. This relationship is illustrated by the following example. A curve with a radius of 2,865 ft would have a degree of curvature of 2°, whereas a curve with a radius of 286.5 ft would have a degree of curvature of 20°.

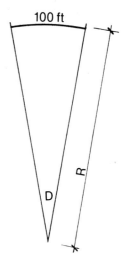

FIGURE 12.5. Degree of Curve.

Example 12.2

Determine the radius, tangent distance, length of curve, and chord length for a 10° curve with an included angle of 60°00'.

Solution. The radius is calculated from the following proportion:

$$R = \frac{5729.578}{D} = \frac{5729.578}{10}$$

$= 572.96$ ft

The remaining information is calculated by the formulas applied in Example 12.1.

Tangent distance

$$T = R \tan \frac{I}{2}$$
$= 572.96 \tan \frac{60°}{2}$
$= 572.96 \tan 30°$
$= 572.96 \times 0.57735 = 330.80$ ft

Length of curve

$$\frac{L}{2\pi R} = \frac{I}{360°}$$

$$L = 2\pi R \times \frac{I}{360}$$
$= 2\pi(572.96) \times \frac{60}{360}$
$= 600.00$ ft

It should be noted that the following proportion, based on degree of curve, may also be used to calculate the length of arc:

$$\frac{L}{100} = \frac{I}{D}$$

$$L = 100 \times \frac{60}{10}$$

$$= 600 \text{ ft}$$

Chord distance

$$C = 2R \sin \frac{I}{2}$$

$$= 2 \times 572.96 \sin \frac{60°}{2}$$

$$= 2 \times 572.96 \sin 30°$$

$$= 2 \times 572.96 \times 0.5000$$

$$= 572.96 \text{ ft}$$

Example 12.3

Determine the degree of curve for Example 12.1.

Solution. The degree of curve is calculated from the proportion:

$$R = \frac{5729.578}{D}$$

The proportion may be arranged as follows:

$$D = \frac{5729.578}{R}$$

As given in the problem, R equals 100 ft; therefore,

$$D = \frac{5729.578}{100}$$

$$= 57.29578°$$

$$= 57°17'45'', \text{ or about } 57°18'$$

STATIONING

Stationing is a measurement convention applied to route surveying for streets, power lines, sanitary and storm sewers, and so on. Stationing is marked out continuously along a route center line, usually at 100-ft intervals, from a starting point designated as station 0+00. In addition, critical points, such as high and low points, street intersections, and beginnings and ends of curves, are located by station points. The typical manner for noting station points is illustrated in Fig. 12.6. As shown in the figure, a separate stationing system is used for each road.

Example 12.4

This example illustrates the procedure for stationing roads that contain horizontal circular curves. Using Example

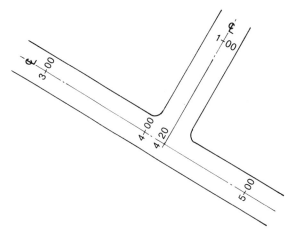

FIGURE 12.6. Typical Center Line Stationing.

12.1, locate the 100-ft stations, the stations for PC and PT, and the station at the end of the road.

Solution. Step 1 The first step in the process is to determine the bearings and lengths of the tangent lines. In addition to the bearings that were previously established in Example 12.1, lengths have now been assigned to tangent line AB (372.25 ft) and tangent line BC (326.90 ft). This information is summarized in Fig. 12.7. Note that point A is the beginning of the road and point C is the end of the road.

Step 2 The next step is to establish the horizontal curves and calculate all necessary horizontal curve data (R, T, L, and I). As previously calculated, the values are as follows:

$I = 88°10'$

$R = 100.00$ ft

$T = 96.85$ ft

$L = 153.88$ ft

Step 3 The third step is to calculate the station points for PC and PT (Fig. 12.8). PC is located by subtracting the horizontal curve tangent length T from the total length of tangent line AB.

```
  372.25 ft
-  96.85 ft
  275.40 ft = station 2+75.40 PC
```

The station at PT is determined by adding the length of the arc L to the previously determined station of PC.

```
  275.40 ft
+ 153.88 ft
  429.28 ft = station 4+ 29.28 PT
```

180 Horizontal Road Alignment

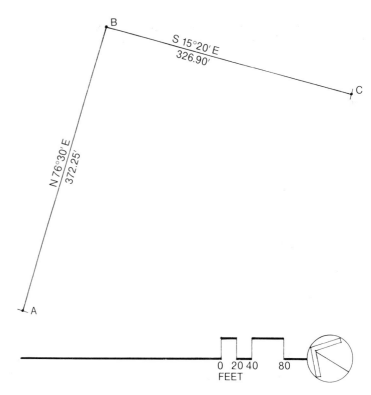

FIGURE 12.7. Tangent Line Lengths and Bearings for Example 12.4.

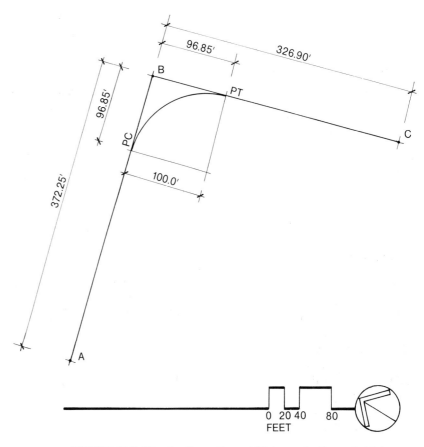

FIGURE 12.8. Circular Curve Tangent Distances for Example 12.4.

Step 4 The last step is to station the tangents and circular curves completely, including 100-ft station intervals and the station at the end of the road (Fig. 12.9). The distance along tangent line *AB* to PC is 275.40 ft; therefore, stations 1+00 and 2+00 must occur along this length. Since the station at PT is 4+29.28, stations 3+00 and 4+00 must occur along the arc. These stations can be located by using the proportion

$$\frac{L}{2\pi R} = \frac{I}{360}$$

to determine the angle formed by the known length of the arc. To locate station 3+00, the length of arc from PC (station 2+75.40) is determined.

$$\begin{array}{r}3+00.00\\-2+75.40\\\hline L = \ 24.60 \text{ ft}\end{array}$$

The angle subtending this length can now be calculated by substituting *L* into the proportion.

$$\frac{I}{360} = \frac{24.60}{2\pi 100}$$

$$I = \frac{24.60}{2\pi 100} \times 360$$

$$= 14.095$$

$$= 14°5'41'' \text{ (or approximately } 14°06')$$

Marking off an angle of 14°06' at the center of the curve from PC will locate station 3+00.

To locate station 4+00, substitute 100 ft for *L*, since stations 3+00 and 4+00 are 100 ft apart.

$$\frac{I}{360} = \frac{100}{2\pi 100}$$

$$I = \frac{100}{2\pi 100} \times 360$$

$$= 57.30$$

$$= 57°18'$$

Marking off an angle of 57°18' the center of the curve from station 3+00 will locate station 4+00. Note that, consistent with the degree of curve definition, the included angle for the 100-ft arc agrees with the degree of curve previously calculated in Example 12.3.

Finally, the end of road station is calculated by determining the length of the tangent line from PT to point *C*. This distance can be determined by subtracting the horizontal curve tangent length *T* from the total length of tangent line *BC*.

$$\begin{array}{r}326.90\\-\ 96.85 \ (T)\\\hline 230.05 \text{ ft}\end{array}$$

This distance is then added to the station at PT.

$$\begin{array}{r}429.28 \text{ ft}\\+230.05 \text{ ft}\\\hline 659.33 \text{ ft} = \text{station } 6+59.33 \text{ (end of road)}\end{array}$$

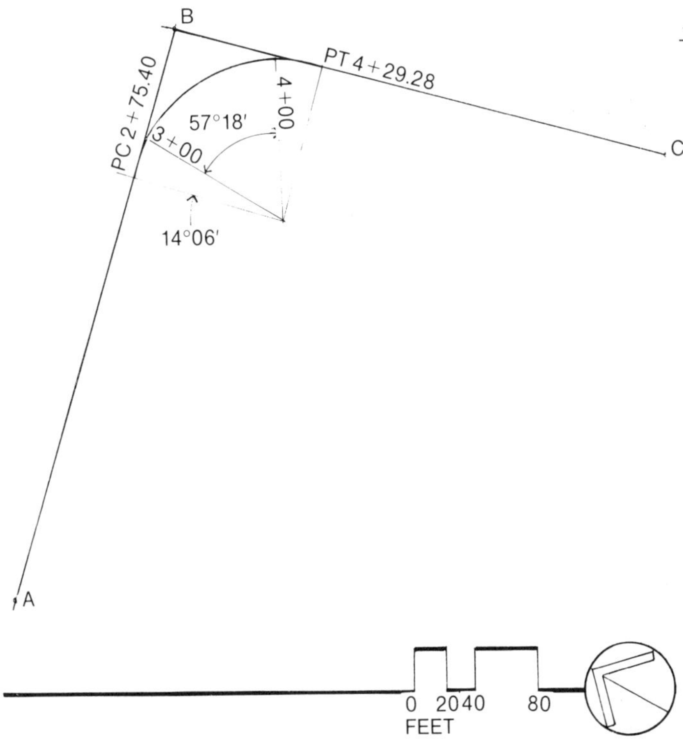

FIGURE 12.9. Curve Stationing for Example 12.4.

Thus the total length of this road is 659.33 ft (Fig. 12.10).

For alignments containing more than one horizontal curve, the procedure is the same. However, once the stations have been determined for the first curve, the next step is to determine the tangent distance from the PT of the first curve to the PC of the second curve (Fig. 12.11). This distance is then added to the station of the first PT to determine the station at the second PC. This process is applied to each successive curve.

It must be emphasized that stationing and curve data are usually determined for the center line of a route alignment.

HORIZONTAL SIGHT DISTANCE

A visual obstruction close to the inside edge of a horizontal curve restricts the driver's view of the road ahead, as illustrated in Fig. 12.12. Preferably, the forward sight distance should not be less than safe stopping distance for the design speed of the curve, thus providing the driver sufficient time to stop once an object has been spotted in the roadway. If the curve is drawn to scale and the obstruction properly located, the sight distance can be obtained by scaling. Sight distances may also be determined analytically by equation, but this is not within the scope of this text.

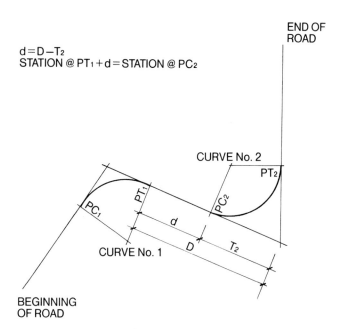

FIGURE 12.11. Determining Stationing for Alignments with More Than One Horizontal Curve.

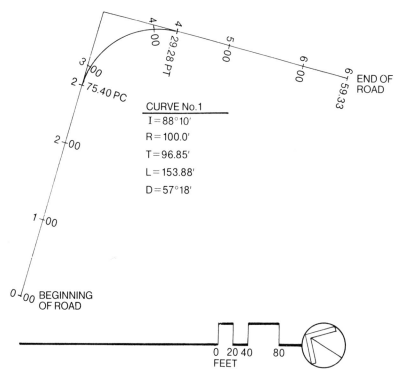

FIGURE 12.10. Complete Alignment Stationing and Circular Curve Data for Example 12.4. Circular curve data are normally shown directly on construction plans as indicated.

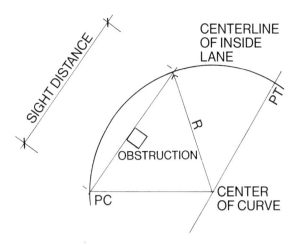

FIGURE 12.12. Horizontal Sight Distance.

CONSTRUCTION DRAWING GRAPHICS

For identification purposes, horizontal curves are usually assigned numbers on a construction drawing. The information presented on drawings includes the following:

1. Included angle (I)
2. Radius (R)
3. Tangent distance (T)
4. Length of curve (L)
5. Degree of curve (D) (optional)
6. Station points for PC and PT
7. Center line stationing
8. Bearings for tangent lines
9. Length and bearing of chord (C) (optional)

The presentation of the information may be arranged in a variety of formats. One typical format is illustrated in Fig. 12.10. Where the scale or complexity of the drawing makes this format difficult, the curve data may be summarized in separate charts or tables.

HORIZONTAL ALIGNMENT PROCEDURE

To this point the problems presented in this chapter have been structured, since specific circular curve data have been predetermined. However, there has been no discussion of the initial procedures involved in establishing the horizontal alignment of a road or path.

Step 1

The first step in the process is to establish freehand *desire lines* based on an analysis of natural and cultural conditions as well as the criteria to be used for the road design on the site plan. Desire lines indicate the optimal path of travel according to the analysis of the site and function of the road and are typically drawn as tangent lines (Fig. 12.13). As mentioned at the beginning of the chapter, an in-depth discussion of these considerations is beyond the intent of this book. However, a brief list of the items of information usually required is presented here for awareness and reference purposes.

Natural site conditions, including topography, soils, vegetation, drainage patterns, and wildlife habitats, should be inventoried and analyzed. In addition, cultural considerations, including neighborhood context, existing traffic

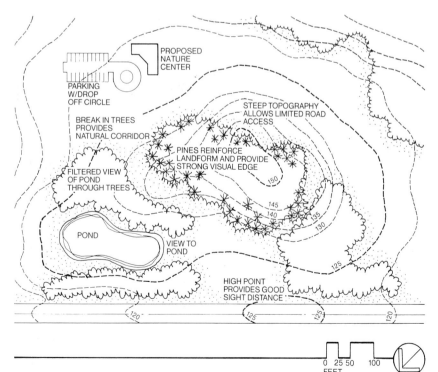

a. Conduct site analysis to determine best location for proposed road or drive.

FIGURE 12.13. Horizontal Alignment Procedure.

FIGURE 12.13. *continued.*

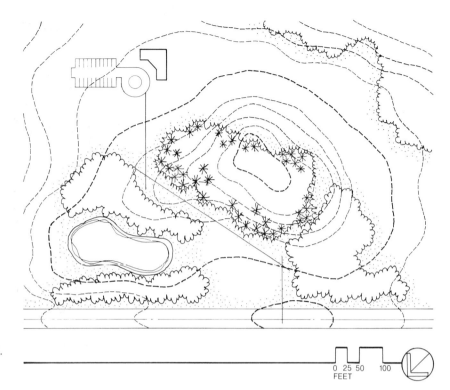

b. Draw in lines along the desired path of travel. These become the tangent lines for the proposed alignment.

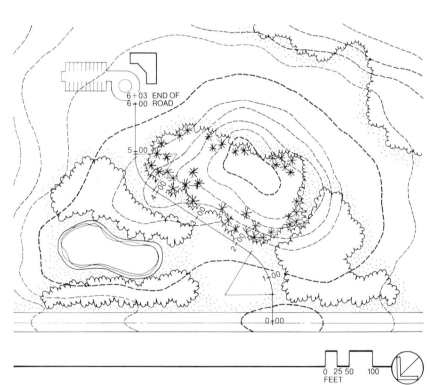

c. Draw in curves and station the road and drive. Usually the curves are first drawn freehand, then circular curves are designed to approximate the desired curves.

patterns, views both to and from the proposed alignment, safety, and potential air and noise pollution, should be analyzed. To establish physical design standards, such as minimum horizontal curve radius, maximum slopes, sight and stopping distances, number of traffic lanes, and cross-sectional design, a profile of the proposed type of use of the proposed alignment must be developed. This includes design speed, types of vehicle, estimated traffic volumes, direction of flow, and hourly and seasonal timing of flow. An additional consideration for any proposed alignment is cost.

Step 2

The next step is to transform the freehand drawing into a preliminary alignment for the road center line. This is done by mechanically drafting the tangents and horizontal curves.

Step 3

Once the center line has been drafted, horizontal curve data can be computed. To do the calculations, however, two values of the circular curve equation must be known. Typically, the deflection angle is measured or determined by bearings if possible. Then *either* the tangent distance or the radius is determined by scaling the drawing (or lengths are assigned). *The deflection angle and the tangent length or radius are the only two quantities that are predetermined or measured.* All other information must be calculated.

Step 4

The last step is to check the horizontal curve calculations against the design criteria and review the entire alignment with regard to the site analysis. Where problems arise, the alignment should be reworked. Finally, the horizontal alignment should be completely stationed.

It should be noted that the formulas and procedures described in this chapter for circular road curves can also be applied to laying out any other circular site feature, such as a path, wall, or fence.

SUPERELEVATION

When a vehicle travels around a curve, a centrifugal force acts on it. To counteract this force to some extent, road surfaces are usually banked or tilted inward toward the center of the curve. The banking of horizontal curves, called *superelevation,* is generally accomplished by rotating the road surface about its center line or, in cut situations, preferably about its inside edge (Fig. 12.14).

Superelevation does not occur abruptly at the PC or PT; rather there is a gradual change as the curve is approached along the tangent lines. The length over which this gradual change occurs is referred to as the *runoff distance,* which is sometimes computed by adding the rate of crown of the road in inches per foot (in./ft) to the rate of superelevation in in./ft and multiplying the sum by 160.

Superelevation can be determined by the formula

$$S = 0.067 \frac{V^2}{R} \qquad (12.5)$$

where: S = superelevation in ft/ft of pavement width
V = design speed, mph
R = radius of curve, ft

Where snow and ice are considerations, S should not exceed 1 in./ft or 0.083 ft/ft. If snow and ice are not problems, a preferred maximum value is 1.5 in./ft or 0.125 ft/ft.

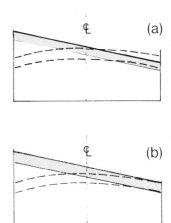

FIGURE 12.14. Superelevation.
a. Rotated about the center line.
b. Rotated about the inside edge.

TABLE 12.1. Alignment Standards in Relation to Design Speed

Design speed (mph)	Minimum radius of horizontal curves (ft)	Maximum percentage of grade	Minimum length of vertical curve for each 1% of algebraic difference (ft)
20	100	12	10
30	250	10	20
40	450	8	35
50	750	7	70
60	1100	5	150
70	1600	4	200

^aAdapted from Lynch (1971)

Example 12.5

Determine the superelevation and runoff distances for a 400-ft radius curve on a road with a 30-mph design speed in a snowy climate. The road has a crown of 0.25 in./ft.

Solution. The superelevation is calculated by substituting into the equation:

$$S = 0.067 \frac{V^2}{R}$$

$$= 0.067 \left(\frac{30^2}{400}\right)$$

$$= 0.15 \text{ ft/ft of width or } 1.8 \text{ in./ft of width}$$

This value exceeds the recommended maximum of 1.0 in./ft; therefore, the proposed superelevation for this curve is 1.0 in./ft, or 0.083 ft/ft.

The runoff distance is calculated by adding the crown rate (0.25 in./ft) and the superelevation rate (1 in./ft) and multiplying by the constant (160).

$$(0.25 + 1.0) \times 160 = 200 \text{ ft}$$

This means that the banking of the road begins at a gradual rate 200 ft before the PC and PT. A technique for handling the transition from the total crown cross section to the total superelevation cross section is illustrated in Fig. 12.15. The full rate of superelevation occurs at the PC and PT and is maintained through the entire curve. As a safety factor, roads are usually widened on the inside of horizontal curves.

EXERCISES

12.1 The center line of a proposed road runs on a direction of N15°00'W to a point A and then changes direction to N58°40'E. A circular curve with a radius of 500.0 ft is to be designed to accommodate the change in direction. Calculate the included angle, tangent distance, curve length, and degree of curve.

12.2 Bearings and tangent lengths for a proposed road alignment are given in Fig. 12.16. Design curve no. 1 with $T = 190$ ft and curve no. 2 with $R = 130$ ft. Provide all horizontal curve data for each curve, including radius, tangent distance, included angle, curve length, degree of curve, and chord length. Station the road completely with 100-ft stations, at points of curvature and tangency, and at the end of the road.

12.3 Compute the superelevation and runoff distances for a 450-ft-radius curve for a road with a 25-mph design speed. The road crown is 1/8 in./ft.

12.4 For an 8° curve with a included angle of 50°, determine the radius, tangent distance, length of curve, and length of chord.

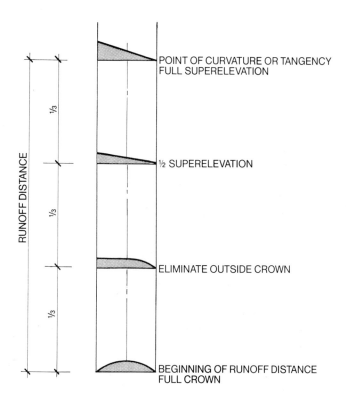

FIGURE 12.15. Transition to Complete Superelevation along Runoff Distance.

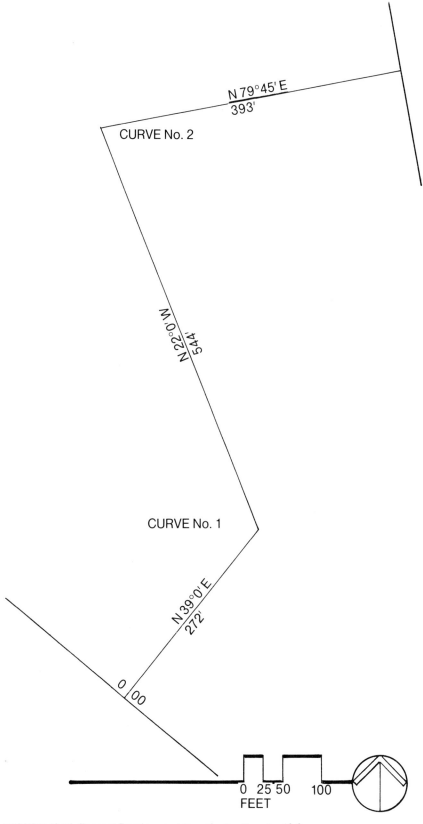

FIGURE 12.16. Tangent Bearings and Lengths for Exercise 12.2.

13
Vertical Road Alignment

Vertical curves are used to ease the transition whenever there is a vertical change in direction (slope). Such transitions eliminate awkward bumps along the vehicular path, allow for proper sight distances, and prevent scraping of cars and trucks on the pavement at steep service drives and driveway entrances. Generally, vertical curves are required for low-speed roads and drives when the vertical change exceeds 1%. This vertical change is computed by determining the algebraic difference between the tangent gradients, with tangents in the uphill direction assigned positive values and those in the downhill direction negative values. Thus, the vertical change for a $+2.0\%$ gradient intersecting with a -2.0% gradient is 4.0% $[+2.0 - (-2.0) = 4.0]$.

There are six possible variations for grade alignment changes. The first is the *peak curve*, in which the entering tangent gradient is positive and the exiting tangent gradient is negative in the direction of stationing. The resultant curve profile is convex. The second is the *sag curve*, with the entering gradient negative and the exiting gradient positive. The resultant curve profile is concave. The four remaining variations are *intermediate peak or sag curves* in which the change in slope occurs in the same direction (i.e., both values either positive or negative), as illustrated in Fig. 13.1.

It is important to note that the computation of horizontal and vertical alignments for roads involves *separate* procedures; thus horizontal and vertical curves may partially or completely overlap or may not overlap at all. Certain relationships between the two curves affect safety and perception of motion; for example, sharp horizontal curves should be avoided at the apex of peak vertical curves. A more thorough discussion of these relationships should be pursued in other texts (see Bibliography). Also, do not

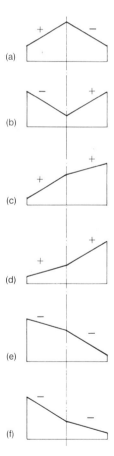

FIGURE 13.1. Vertical Curve Tangent Variations.
a. Peak curve.
b. Sag curve.
c. Intermediate peak curve.
d. Intermediate sag curve.
e. Intermediate peak curve.
f. Intermediate sag curve.

become confused by the similarity in terminology for horizontal and vertical curves. Realize that tangent lines for horizontal curves are *direction* lines in the horizontal plane (*through* the landscape), whereas tangent lines for vertical curves are *slope* lines in the vertical plane (*over* the landscape).

VERTICAL CURVE FORMULA

Generally, parabolas are used for vertical curves, since they lend themselves to simple computation and leveling procedures. For parabolas, the offset distances from the tangent line to the curve vary as the square of the distances from either end of the curve (see Fig. 13.2).

The components of vertical curves are illustrated in Fig. 13.2. Note that the illustration depicts a peak curve. There are two types of vertical curves: equal tangent curves and unequal tangent curves. For *equal tangent curves*, also referred to as symmetrical, the horizontal distance from the beginning of curve (BVC) to the point of vertical intersection (PVI) *equals* the horizontal distance from the PVI to the end of curve (EVC). For *unequal tangent curves*, also called asymmetrical, the horizontal distance from BVC to PVI does not equal the horizontal distance from PVI to EVC.

UNEQUAL TANGENT CURVES

The formulas for computing asymmetrical curves are

$$e = \frac{l_1 l_2}{200(l_1 + l_2)} \times A \qquad (13.1)$$

where: e = tangent offset at PVI, ft
l_1 = horizontal distance from BVC to PVI, ft
l_2 = horizontal distance from EVC to PVI, ft
A = algebraic differences between tangent gradients, %

and

$$y_1 = e\left(\frac{x_1}{l_1}\right)^2; \qquad y_2 = e\left(\frac{x_2}{l_2}\right)^2 \qquad (13.2)$$

where: y_1 = vertical distance from entering tangent line to curve (tangent offset), ft
x_1 = horizontal distance from BVC to point on curve, ft
y_2 = vertical distance from exiting tangent line to curve (tangent offset), ft
x_2 = horizontal distance from EVC to point on curve, ft
l_1, l_2, and e have previously been defined.

Example 13.1

For the preliminary profile for a road center line, a +2.0% grade intersects a −3.0% grade at station 35+40. The elevation at the PVI is 261.40 ft. The horizontal length of the entering tangent is 200 ft; the length of the exiting tangent is 300 ft. Calculate the elevations of the curve at all 100-ft stations (Fig. 13.3).

Solution. First the tangent offset at the PVI must be calculated.

$$\begin{aligned}
e &= \frac{l_1 l_2}{200(l_1 + l_2)} \times A \\
&= \frac{200 \times 300}{200(200 + 300)} \times 5 \\
&= \frac{60{,}000}{100{,}000} \times 5 \\
&= \frac{300{,}000}{100{,}000} = 3
\end{aligned}$$

Next the elevations of the 100-ft stations along the tangent lines and at the BVC and EVC are calculated.

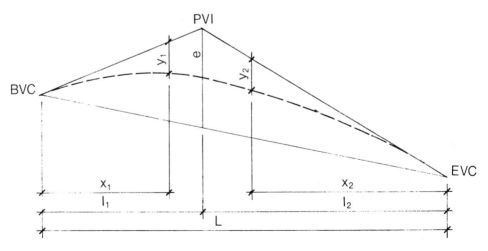

FIGURE 13.2. Elements of Asymmetrical Vertical Curves.

Station 33+40 (BVC): 261.40 − (200 × .02) = 257.40
34+00   261.40 − (140 × .02) = 258.60
35+00   261.40 − (40 × .02) = 260.60
35+40 (PVI)   = 261.40
36+00   261.40 − (60 × .03) = 259.60
37+00   261.40 − (160 × .03) = 256.60
38+00   261.40 − (260 × .03) = 253.60
38+40 (EVC): 261.40 − (300 × .03) = 252.40

The tangent offset distances are then calculated for the entering tangent and exiting tangent; they are based on the horizontal distance from BVC and EVC, respectively.

Entering tangent

$$y = e\left(\frac{x_1}{l_1}\right)^2 \tag{13.3}$$

Station 34+00: $y = 3\left(\frac{60}{200}\right)^2 = 0.27$ ft

Station 35+00: $y = 3\left(\frac{160}{200}\right)^2 = 1.92$ ft

Exiting tangent

Station 36+00: $y = 3\left(\frac{240}{300}\right)^2 = 1.92$ ft

Station 37+00: $y = 3\left(\frac{140}{300}\right)^2 = 0.65$ ft

Station 38+00: $y = 3\left(\frac{40}{300}\right)^2 = 0.05$ ft

Finally, the tangent offset distances are *subtracted* from the tangent elevations to determine the curve elevations, since this is a peak curve. Table 13.1 illustrates a typical form used to record curve data.

EQUAL TANGENT CURVES

Most vertical curves are designed as equal tangent or symmetrical curves. This simplifies the calculations, since l_1 equals l_2. Thus the previous equations can be simplified to one generalized equation:

$$y = e\left(\frac{x}{l}\right)^2 \tag{13.4}$$

where: y = tangent offset distance, ft
x = horizontal distance from BVC (or EVC) to point on curve, ft
l = one-half length of curve, ft
e = tangent offset at PVI, ft

Points located at the same horizontal distance from the BVC and EVC have the same tangent offset distance, but not necessarily the same curve elevation. This is demonstrated in the following example.

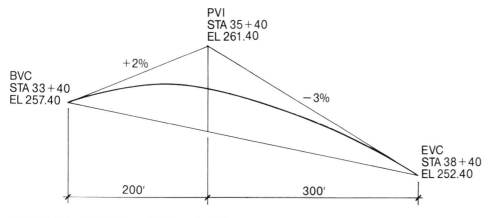

FIGURE 13.3. Vertical Curve for Example 13.1.

TABLE 13.1. Vertical Curve Data

Station	Point	Tangent elevation	Tangent offset	Curve elevation
33+40	BVC	257.40	0.00	257.40
34+00		258.60	0.27	258.33
35+00		260.60	1.92	258.68
35+40	PVI	261.40	3.00	258.40
36+00		259.60	1.92	257.68
37+00		256.60	0.65	255.95
38+00		253.60	0.05	253.55
38+40	BVC	252.40	0.00	252.40

Example 13.2

On a preliminary profile of the center line of a road, a -2.0% grade intersects at $+1.5\%$ grade at station $79+00$.

The elevation at the PVI is 123.50 ft. A vertical curve with $L = 800$ ft is desired. Calculate the elevations of the curve at all 100-ft stations (Fig. 13.4).

Solution. The first step is to calculate the elevations at BVC and EVC and at 100-ft intervals along both tangent lines. Since the curve is symmetrical about the PVI, the station at BVC is $75+00$ and the station at EVC is $83+00$. Elevations at these points can now be determined by applying the slope formula.

$(400 \times 0.02) + 123.50 = 131.50$ elevation at BVC
$(400 \times 0.015) + 123.50 = 129.50$ elevation at EVC

The elevations at 100-ft intervals along the tangent lines are easily determined, since, for the entering tangent, a -2.0% gradient produces a 2.0-ft drop in elevation for every 100 ft of distance from the BVC. For the exiting tangent, the change in elevation is 1.5 ft for every 100 ft. The tangent elevations at the 100-ft stations are summarized in Table 13.2.

The next step is to determine the height of the middle offset. From the previous equation simplified, the height is computed as follows:

$$e = \frac{l^2}{200 \times 2 \times l} \times A \qquad (13.5)$$

$$= \frac{400^2}{200 \times 2 \times 400} \times 3.5 = 3.50 \text{ ft}$$

The elevation of the curve at the PVI station $79+00$ is $123.50 + 3.50 = 127.0$ ft.

With e known, the tangent offsets can be calculated for the desired 100-ft intervals.

Station $75+00$: $y_0 = 3.5 \, (\frac{0}{400})^2 = 0.00$ ft

$76+00$: $y_1 = 3.5 \, (\frac{100}{400})^2 = 0.22$ ft

$77+00$: $y_2 = 3.5 \, (\frac{200}{400})^2 = 0.88$ ft

$78+00$: $y_3 = 3.5 \, (\frac{300}{400})^2 = 1.97$ ft

$79+00$: $y_4 = 3.5 \, (\frac{400}{400})^2 = 3.50$ ft

Station $79+00$ is the center of the curve.

Since both tangent lines are the same horizontal length from the ends of the curve to the PVI, the tangent offsets are symmetrical about the PVI and do not have to be computed again for the curve from PVI to EVC.

Since this is a sag curve, the tangent offset distances are added to the corresponding tangent elevations to determine the elevations of the curve. The resultant vertical curve is plotted in Fig. 13.5. For peak curves, the tangent offset distances would be subtracted from the tangent elevations. The data for the curve are summarized in Table 13.2. Table 13.2 also demonstrates a convenient check for vertical curve computations. The *second differences* of the tangent offsets at *equal horizontal intervals* are constant. (The discrepancy between 0.44 and 0.43 results from rounding.)

CALCULATION OF LOCATIONS OF HIGH AND LOW POINTS

The locations of low points are needed to position drainage structures; those of high points may be required to determine sight distances. Locating both high and low points may be necessary to determine critical clearances under structures such as bridges.

The high or low point of a vertical curve coincides with the midpoint of an equal tangent curve only when the gradients of the tangent lines are equal. In all other cases, the high or low point is located on the side of the PVI *opposite* the steepest gradient.

The formula for locating the high or low point of a vertical curve is

$$d = \frac{L g_1}{(g_1 - g_2)} \qquad (13.6)$$

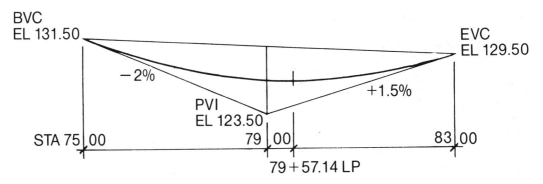

FIGURE 13.4. Vertical Curve for Example 13.2.

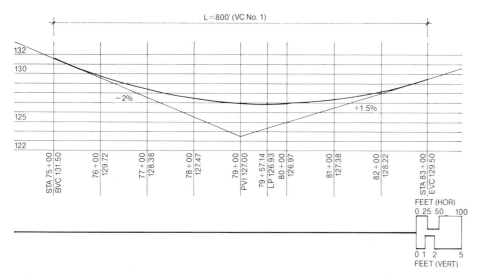

FIGURE 13.5. Vertical Curve Profile.

TABLE 13.2. Vertical Curve Data

Station	Point	Tangent elevation	Curve elevation	Tangent offset	First difference	Second difference
75+00	BVC	131.50	131.50	0.00		
76+00		129.50	129.72	0.22	0.22	0.44
77+00		127.50	128.38	0.88	0.66	0.43[a]
78+00		125.50	127.47	1.97	1.09	0.44
79+00	PVI	123.50	127.00	3.50	1.53	
79+57.14	LP	124.36	126.93	2.57		
80+00		125.00	126.97	1.97		
81+00		126.50	127.38	0.88		
82+00		128.00	128.22	0.22		
83+00	EVC	129.50	129.50	0.00		

[a] Discrepancy due to rounding.

where: d = distance from BVC to HP or LP, ft
L = total length of curve, ft
g_1 = gradient from BVC to PVI
g_2 = gradient from PVI to EVC

Example 13.3

A drain inlet is to be installed at the low point of the curve in the previous example. Find its location and elevation.

Solution. To determine the distance from BVC, substitute the known values into the equation.

$$d = \frac{800 \times (-0.02)}{-0.02 - (+0.015)}$$

$$= \frac{-16.0}{-0.035} = 457.14 \text{ ft}$$

Thus, the low point occurs 457.14 ft from BVC, which is station 79+57.14.

Next, compute the tangent elevation at station 79+57.14. This location is 57.14 ft past the PVI; therefore, the tangent elevation is 123.50 + (57.14 × 0.015) = 124.36 ft. The distance from EVC is 400.00 − 57.14 = 342.86 ft, and the tangent offset is determined as

$$y_{lp} = 3.5 \left(\frac{342.86}{400}\right)^2 = 2.57 \text{ ft}$$

The elevation of the low point on the curve is 124.36 + 2.57 = 126.93 ft. Note that this elevation is for the center line of the road. If the drain inlet is located at the edge of the road, an adjustment based on the road cross section would be necessary to determine the top of frame elevation. For example, if the crown height of the proposed road is 0.25 ft., then the elevation for the drain is 126.93 − 0.25 = 126.68 ft.

CONSTRUCTION DRAWING GRAPHICS

Unlike horizontal curves, for which all data are presented on the layout plan, vertical curve data are presented on a profile of the road center line. The presentation format for the curve in Example 13.2 is shown in Fig. 13.5. Information provided on profiles includes

1. Vertical curve number (for identification purposes)
2. Total length of curve L
3. Stationing at BVC, PVI, EVC, HP or LP, and 100-ft intervals
4. Curve elevations for all stations in (3)
5. Tangent gradients

Figure 13.5 indicates the profile of only one curve. However, the entire road from station 0+00 to the end would normally be profiled. The profile of the edge of the road may also be indicated by a dashed line on the same drawing.

It should be noted that the *horizontal* alignment must be calculated before the profile can be constructed, since stationing occurs along the center line of horizontal curves. A profile is then constructed from the beginning to the end of the road and represents the total length along the curved as well as the straight portions.

VERTICAL SIGHT DISTANCES

There are two types of sight distance. The first is safe stopping sight distance; in other words, the distance required to react, brake, and stop a vehicle at a given speed. The second is safe passing sight distance. Both of these are a concern on peak curves, since the convex profile shortens the line of sight. For low-speed roads in which landscape architects are involved, passing will most likely be prohibited; therefore, safe stopping distance is of greater concern. Roads requiring safe passing distances should be designed by qualified highway engineers.

In determining safe stopping sight distance, generally an eye height of 3.75 ft and object height of 0.50 ft are used. Although formulas may be used to calculate the minimum length of vertical curve necessary to maintain a safe stopping distance at a given speed, sight distances for vertical curves can be determined to a certain degree by measuring from the height of the eye to the height of the object on the profile as shown in Fig. 13.6. Where the measured distance is less than the safe stopping sight distance, the vertical curve must be redesigned. Again, a highway engineer should be consulted if sight distances are a critical concern.

ROAD ALIGNMENT PROCEDURE

The following outline is a systematic procedure which may be used to ease the task of laying out both horizontal and vertical alignment for roads, drives, and paths.

Step 1

As discussed briefly in Chapter 12, the first steps are to develop design criteria and constraints for both horizontal and vertical curves and to conduct a site analysis to determine the best route or corridor location (see Fig. 12.13a).

Step 2

Once step 1 has been completed, desire lines can be established *through* the landscape. Desire lines represent movement along the horizontal plane; therefore, it is necessary to design horizontal curves to make this movement or flow as smooth as possible. During this step, then, horizontal curves should be preliminarily designed, all necessary data calculated, and the center line of the road or path completely stationed (Figs. 12.13b and 12.13c).

Step 3

Next, a profile of the *existing* grades along the *proposed* center line is constructed (Fig. 13.7).

Step 4

At this point the alignment *over* the landscape can be designed. To begin, the *proposed* vertical curve tangent lines are placed on the profile of existing grades. The actual placement of these tangent lines is influenced by many factors, including balancing of cut and fill, design speed, roughness of topography, and horizontal curve placement (Fig. 13.8).

Step 5

From the profile, station points and elevations for the intersections of the proposed vertical curve tangent lines are established (Fig. 13.9). The differences in elevation between intersection points can now be determined. Since

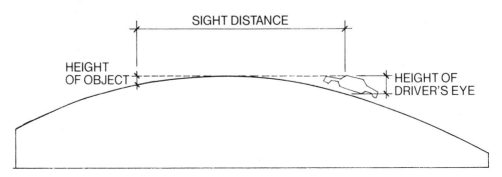

FIGURE 13.6. Sight Distance for Peak Curves.

Road Alignment Procedure 195

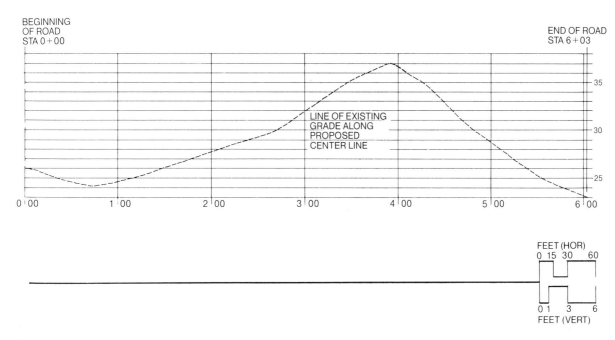

FIGURE 13.7. Profile of Existing Grades along Proposed Center Line.

the distances between the intersection points can be calculated from the stationing, the slopes of the tangent lines can be computed.

Step 6

The next step is to determine the lengths of vertical curves required. This depends on such factors as design speed, topography, aesthetics, and sight distances. Vertical curves may be designed as equal tangent (symmetrical about the PVI) or unequal tangent (asymmetrical about the PVI) curves (Fig. 13.10).

Step 7

At this point the slope of the tangents, lengths of curves, elevations at points of vertical intersection, and stations for all BVCs, PVIs, and EVCs are known. Therefore, all data necessary to calculate the vertical curves are available and the profile can be completed.

Step 8

Once the profile has been completed, the *proposed* grades are transferred from the profile to the center line of the

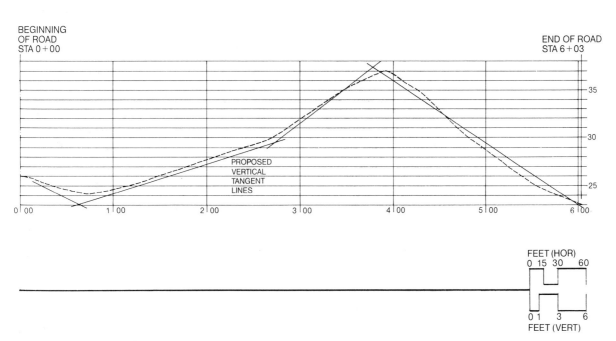

FIGURE 13.8. Proposed Vertical Tangent Lines.

196 Vertical Road Alignment

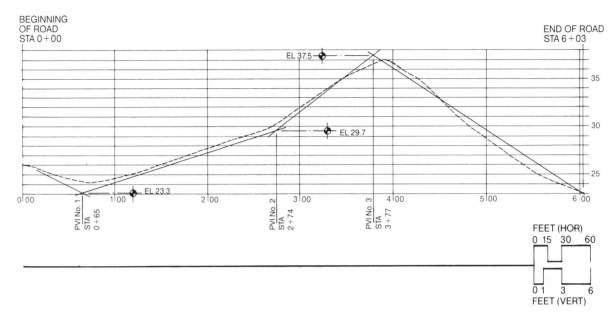

FIGURE 13.9. Stations and Elevations for Proposed Points of Vertical Intersection.

horizontal alignment on the plan. The road (path, etc.) is graded according to the proposed cross-sectional design and the proposed contour lines are appropriately connected to the existing contour lines. The proposed grading plan is now analyzed for problem areas, such as steep slopes, excessive cuts or fills, drainage, and removal of vegetation. Where problems arise, either the vertical alignment (profile) or the horizontal alignment (plan) or both must be restudied and the profile and plan adjusted accordingly.

EXERCISES

13.1 A proposed park road, straight in plan view, crosses a small stream where a vertical sag curve is required. One of the tangents slopes toward the east with a −2.0% grade through station 14+00, which is at elevation 110.0 ft. The other tangent slopes west with a −5.0% grade through station 23+00, at elevation 120.0 ft. Find the location and elevation of the point of intersection of the tangents. Once the PVI has been determined, design

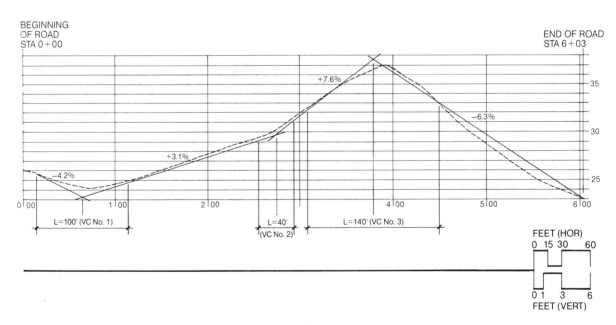

FIGURE 13.10. Tangent Line Slopes and Vertical Curve Lengths.

FIGURE 13.11. Reverse Horizontal Curve with Sag Vertical Curve.

a symmetrical curve with a total length of 400 ft. Determine the elevations of the curve at BVC, EVC, 100-ft stations, and the low point.

13.2 From the information given, calculate the curve elevations for an equal tangent curve at each 50-ft station (0+00, 0+50, etc.) and determine the station points for BVC, EVC, and the high point.

a. Slope of entering tangent: +3.2%
b. Slope of exiting tangent: −4.4%
c. Tangent intersection at station 6+00
d. Elevation at PVI: 25.9 ft
e. Total length of curve: 400 ft

Draw a profile of the curve using a horizontal scale of 1 in. = 50 ft and vertical scale of 1 in. = 5 ft.

13.3 From the information given, calculate the curve elevations for an equal tangent curve at each 50-ft station and determine the station points for BVC, EVC, and the low point.

a. Slope of entering tangent: −6.0%
b. Slope of exiting tangent: +3.8%
c. Tangent intersection at station 5+30
d. Elevation at PVI: 29.2 ft
e. Total length of curve: 300 ft

Again draw a profile of the curve, using the same scales as in Exercise 13.2.

13.4 The plan of a proposed road alignment is shown in Fig. 13.13. Construct a profile of the existing topography along the proposed center line with a horizontal scale of 1 in. = 50 ft and a vertical scale of 1 in. = 5 ft. Design a vertical alignment that works well with the existing topography, assuming a design speed of 20 mph. Transfer the proposed grades from the profile to the plan, grade the road with a 4-in. crown, and connect the proposed to the existing contour lines in an appropriate manner.

FIGURE 13.12. Reverse Horizontal Curve with Peak Vertical Curve. Curving wall reinforces horizontal alignment; constant wall height emphasizes vertical alignment.

198 Vertical Road Alignment

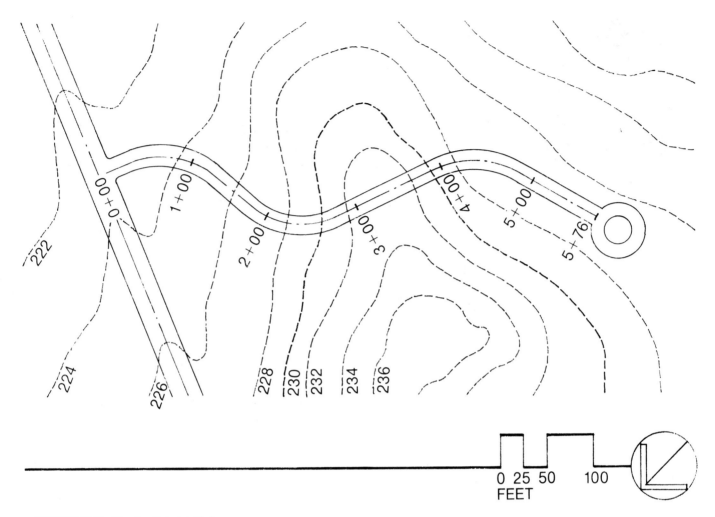

FIGURE 13.13. Plan for Exercise 13.4.

Epilogue

To this point in the text, the components of site engineering have been presented for the most part as independent problems which have been solved separately. However, it is essential to understand that grading, storm water management, earthwork, and road and path alignment are *dependent* problems which are solved holistically in the context of a design program *and* a thorough understanding of a site's specific natural and cultural characteristics.

A final case study is presented here to reinforce an integrated, comprehensive approach to site planning, design, and engineering. The examples in Chapter 6, although primarily selected to illustrate the importance of grading as a design element, also demonstrate a comprehensive design approach.

MORRIS ARBORETUM

Originally a private estate, the 175-acre Morris Arboretum in Philadelphia, Pennsylvania, lacked the necessary physical infrastructure to fulfill its function as a cultural and university institution. In 1976, under the leadership of Dr. William M. Klein, the arboretum hired the ecological planning and design firm of Andropogon Associates, Ltd., to develop a master plan. Part of the challenge of the master plan was to tie together the various landscapes and landholdings which the arboretum comprises and to develop a meaningful entry and arrival experience to the arboretum (Fig. E.1).

Entrance Road

The new entrance on Northwestern Avenue (which bisects the arboretum) is designed to connect the natural, park, and garden landscapes of the former estate on one side of the street with the working landscape of the maintenance, nursery, and research centers on the other. The location and alignment of the road address concerns (traffic, glare from headlights and night lighting, views to and from the site, neighborhood character, etc.) expressed by adjacent residents and respond to physical limitations and opportunities on the site. The alignment consists of a long sweeping curve across the lower meadow and reverse curves (switchbacks, which are necessary to lessen the road gradient) on the steep hillside leading to the education center and parking area.

Two different techniques are used to handle storm runoff from the road. Granite cobble gutters are installed along the steeper portions of the road to collect and conduct runoff to surface inlets so that potential erosion problems are averted. From the surface inlets the water is directed to recharge trenches, where it infiltrates the soil and replenishes the groundwater. By eliminating curbs or gutters on the flatter portions through the lower meadow (Fig. E.2), sheet flow from the road runoff is continued and the problems associated with concentrated runoff are prevented (Fig. E.3).

The two techniques also reinforce the landscape character of the arrival sequence. The absence of curbs preserves the image of a country road and pastoral landscape. The use of the cobble gutters projects a villagelike appearance as the more intensely developed portion of the site is approached (Fig. E.4).

Parking Area

The parking area is placed on top of the hill, facilitating convenient access to the education center and the gardens of the former estate and providing views of the lower

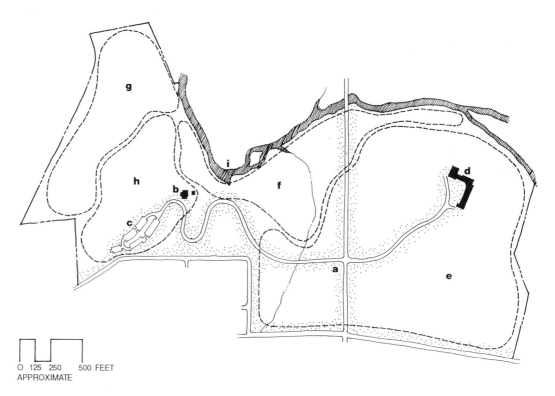

FIGURE E.1. Schematic Plan of the Morris Arboretum. a. Northwestern Avenue entrance. b. Widener Education Center. c. Parking area. d. Horticultural Center. e. Working landscape. f. Natural landscape. g. Park landscape. h. Garden landscape. i. Wissahickon Creek.

FIGURE E.2. Long, Sweeping Curve through Lower Meadow. Eliminating curbs or gutters enhances pastoral character of road.

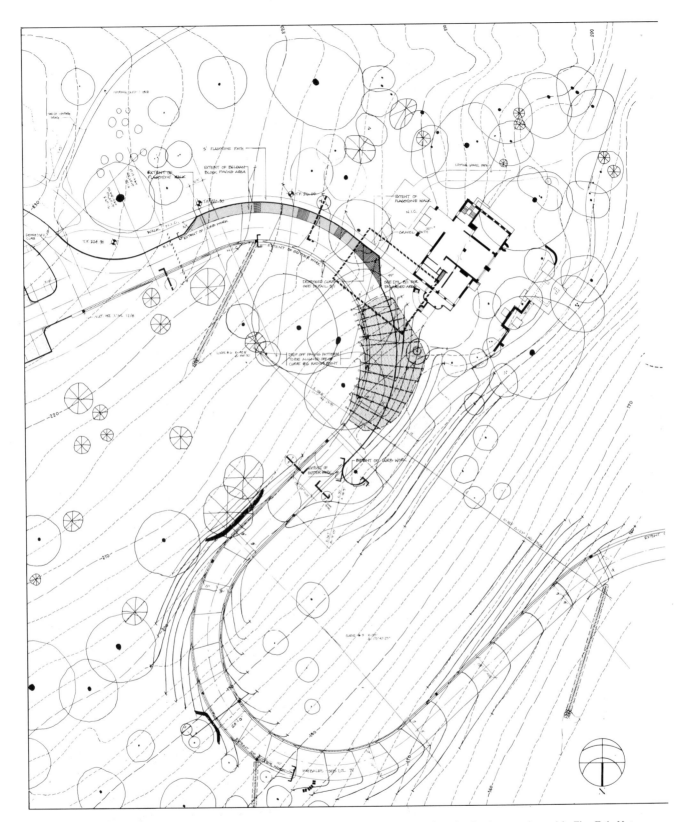

FIGURE E.3. Partial Plan of Road Grading. This drawing is the actual grading construction plan for the area pictured in Fig. E.4. Note configuration of the contour lines as the road cross section changes from crowned to cross-sloped.

FIGURE E.4. Reverse Curve across Steep Slope. Cobble gutters are used to collect and conduct runoff to prevent erosion. Note that the curves have been cross-sloped to create a superelevated condition.

(a)

meadow, creek, and working landscape. The parking area is terraced to fit the topography of the hilltop. The slope separating the two levels of parking helps visually to reduce the apparent scale of the parking area while providing a substantial area for planting.

Management of storm water runoff from the parking area is accomplished by the use of porous asphalt pavement and subsurface stone recharge beds. Porous asphalt pavement is used for the parking bay areas only; standard asphalt pavement is used for the aisles and entrance road (Fig. E.5). This technique recognizes the reduced strength of porous pavement and minimizes potential problems caused by traffic use. Damage to porous paving, if it occurs, appears to result from shear stress caused by the acceleration or braking of vehicles and causes surface scuffing rather than structural cracking. Storm water rapidly percolates through the porous asphalt to the recharge reservoirs beneath the pavement, where water is held in the pore space of the open graded aggregate and slowly infiltrates the soil. At the Morris Arboretum these recharge beds are designed to hold the 100-yr storm and are 18- to 24-in. deep. A recharge bed, unlike a retention basin, does not function by holding accumulated runoff until it can be released to a storm drain, but by leaking water continuously into the subsoil to replenish groundwater. With porous asphalt the surface is level, which allows for even infiltration of stormwater. Pipes direct overflow water to recharge trenches and surface outlets (see Fig. 7.8).

It must be stressed that the design of porous asphalt pavement systems must be site-specific and that such systems are not applicable to every site. Constraints and limitations include soil infiltration rates, depth to water table and bed-

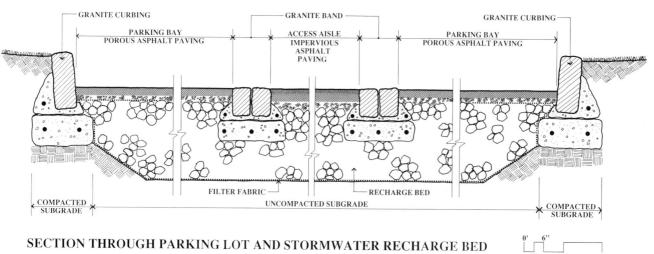

SECTION THROUGH PARKING LOT AND STORMWATER RECHARGE BED

(b)

FIGURE E.5. Parking Area. a. The flush cobble strip separates the standard asphalt pavement of the aisle from the porous asphalt pavement of the parking bays. The darker appearance of the porous pavement is caused by the slightly rougher texture of the larger, open-graded aggregate used in the asphalt mix. b. Section through parking lot illustrating storm water recharge bed and separation of impervious pavement of access aisle from porous pavement of parking bays. *Source:* Andropogon Associates, reprinted with permission.

rock, and traffic type and volume. In particular, porous paving should *not* be installed in areas where the water table is at or near the surface or on limestone substrates where direct contact with water may dissolve the bedrock.

Particular care must be exercised during the construction of porous asphalt pavements: Fine particulates must be prevented from entering and clogging the pavement and recharge basin; heavy equipment must be prevented from compacting the base of the basin; and the filter fabric must be properly installed.

The design of the Morris Arboretum entrance clearly expresses the role of the institution. It takes what would normally be treated as a utilitarian component (i.e., access and parking) and elevates it to one which informs, educates, and contributes to the experience of the place.

Glossary

Abrasion Wearing away by friction.

Alignment The course along which the center line of a roadway or channel is located.

Angle of repose The angle which the sloping face of a bank of loose earth or gravel or other material makes with the horizontal.

Antecedent precipitation Rainfall that has recently preceded the storm event being analyzed.

Area drain A structure for collecting runoff from relatively small paved areas.

Backfill Earth or other material used to replace material removed during construction, such as in pipeline and culvert trenches and behind retaining walls.

Base (Course) A layer of specified or selected material of planned thickness, constructed on the subbase or on the natural subgrade to distribute the load and provide drainage, or on which a wearing surface or a drainage structure is placed.

Bearing capacity (soil) The load-supporting capacity of a soil.

Blading Planing or smoothing the ground surface.

Borrow area A source of earth fill materials used in the construction of embankments or other earth fill structures.

Catch basin A receptacle, with a sediment bowl or sump, for diverting surface water to a subsurface pipe.

Center line The survey line in the center of a road, ditch, or similar project.

Channel A natural stream, or a ditch or swale constructed to convey water.

Compaction The densification of a soil by a mechanical process.

Continuity equation A formula expressing the principle of conservation of mass as applied to the flow of water (or other fluids of constant density), which states that the product of cross section of flow and velocity at any point in a channel is a constant.

Contour line An imaginary line, or its representation on a map, following all points at the same elevation above or below a given datum.

Critical depth The depth of flow in an open channel at which critical flow occurs. For a given flow rate, depths greater than critical result in subcritical, or tranquil, flow. Those smaller than critical result in supercritical, or rapid, flow.

Critical flow An unstable flow condition in open channel which occurs at critical depth.

Critical velocity The velocity of flow in an open channel which occurs at critical depth.

Crown The rise or difference in elevation between the edge and the center line of a roadway.

Culvert Any structure not classified as a bridge which provides a waterway or other opening under a road.

Cut section (or cut) That part of the ground surface which, when graded, is lower than the original ground.

Datum A horizontal reference plane used as a basis for computing elevations.

Detention basin (dry pond) An impoundment, normally dry, for temporarily storing storm runoff from a drainage area to reduce the peak rate of flow.

Discharge q Flow rate in a culvert, pipe, or channel.

Diversion A channel, with or without a supporting ridge on the lower side, constructed across a slope to intercept surface runoff.

Drainage Interception and removal of groundwater or surface water by artificial or natural means.

Drainage area The area drained by a channel or a subsurface drain.

Edaphology The study of the soil from the standpoint of higher plants and crop production.

Elevation (a) The altitude relative to a given datum. (b) A scale drawing of the facade of a structure.

Embankment A bank of earth, rock, or other material constructed above the natural ground surface.

Emergency spillway A channel for safely conveying flood discharges exceeding the capacity of the principal spillway of a detention or retention pond.

Erodibility The susceptibility of soil material to detachment and transportation by running water or wind.

Erosion Detachment and movement of soil or rock fragments by water, wind, ice, and gravity.

Excavation (a) The act of taking out materials. (b) The hollow or depression after the materials have been removed.

Fill section (or fill) That part of the ground surface which, when graded, is higher than the original ground.

Filter strip A vegetated buffer zone for removing sediments and pollutants before runoff reaches ponds, waterways, or other drainage facilities.

Fine grade Preparation of the subgrade preceding placement of surfacing materials.

Foundation The portion of a structure (usually below ground level) that distributes the pressure to the soil or to artificial supports.

Free water Soil water that moves by gravity, in contrast to capillary and hydroscopic water.

French drain A trench filled with coarse aggregate (with or without a pipe) for intercepting and conveying ground water.

Gabion A compartmented rectangular container, made of steel wire mesh and filled with stone; used for erosion control and retaining wall purposes.

Grade

   Finished grade The completed surfaces of lawns, walks, and roads brought to grades as designed.

   Natural grade The undisturbed natural surface of the ground.

   Subgrade The grade established in preparation for top surfacing of roads, lawns, etc.

Gradient The degree of inclination of a surface, road, or pipe, usually expressed as a percentage.

Grading Modification of the ground surface by cuts and/or fills. Fine or finish grading is light or thin grading to finish a prepared earth surface.

Grassed waterway A natural or constructed channel, usually broad and shallow and covered with erosion-resistant vegetation, used to conduct surface runoff.

Gravel Aggregate composed of hard and durable stones or pebbles, crushed or uncrushed, often mixed with sand.

Groundwater Free subsurface water, the top of which is the water table.

Gutter An artificially surfaced and generally shallow waterway, usually provided at the sides of a roadway for carrying surface drainage.

Headwall A vertical wall at the end of a culvert to support the pipe and prevent earth from spilling into the channel.

Hydraulic radius The cross-sectional area flow of a pipe or channel divided by the wetted perimeter.

Hydrograph A graph showing, for a given point on a channel, the discharge, stage, velocity, or other property of water with respect to time.

Hydrologic condition Vegetative cover, residue, and surface roughness of a soil as they may affect potential runoff.

Hydrologic cycle The concept of a closed system, involving the transformation of water from the vapor phase to the liquid (and solid) phase and back to the vapor phase, and the movement of that water.

Hydrologic soil group (HSG) A soil classification system based on infiltration and potential runoff characteristics.

Imperviousness The property of a material through which water will not flow under ordinary hydrostatic pressure.

Infiltration The downward entry of water into the surface of a soil or other material, as contrasted with *percolation,* which is movement of water through soil layers or material.

Infiltration basin An open surface storage area with no outlet, except an emergency spillway, which permits runoff to infiltrate the soil.

Initial abstraction (I_a) Losses before runoff begins, including infiltration, evaporation, interception by vegetation, and water retention in surface depressions.

Inlet An arrangement for conveying surface water to an underdrain.

Intercepting ditch An open drain to prevent surface water from flowing down a slope by conducting it around the slope. See also *diversion.*

Interpolation (topographic) The process of determining the location of elevations from the plotted locations of known elevations.

Invert The lowest point of the internal cross section of a pipe or of a channel.

Low-flow channel A small ditch, constructed in flat bottoms of larger ditches or detention basins, to facilitate their drainage during periods of low flow.

Manhole A structure, covered with a lid, which allows a person to enter a space below ground level.

Manning's equation A formula for calculating the velocity of flow in a channel as a function of relative roughness, cross-sectional configuration, and gradient.

Maximum potential retention (S) The greatest proportion of precipitation that could possibly be retained by a specific soil and land use combination.

Modified Rational method (MRM) An extension of the Rational method for calculating the rate of runoff from a drainage area, including provisions for antecedent precipitation and for development of hydrographs.

Moisture content The percentage, by weight, of water contained in soil or other material, usually based on dry weight.

Monument A boundary stone or other permanent marker locating a property line or corner.

Outlet Point of water disposal from a stream, river, lake, tidewater, or artificial drain.

Peak discharge The maximum instantaneous flow rate resulting from a given storm condition at a specific location.

Percolation Movement of soil water toward the water table.

Perron An exterior platform at a building entrance, usually with steps leading up to it.

Pervious The property of a material which permits movement of water through it under ordinary hydrostatic pressure.

pH A measure of alkalinity or acidity; pH 7 is neutral and pH 6.5 represents a desirable degree of soil acidity.

Porous Having many small openings through which liquids may pass.

Porous pavement A pavement constructed from a material that permits percolation of storm water to the subgrade.

Principal spillway A component of retention or detention ponds, generally constructed of permanent materials. It is designed to regulate the normal water level, provide flood protection, and/or reduce the frequency of operation of the emergency spillway.

Pedology The study of the soil as a natural body, including its origin, characteristics, classification, and description.

Rainfall intensity *(i)* The rate at which rain falls. In the United States usually measured in inches per hour (iph).

Ramp An inclined plane serving as a way between two different levels.

Rational method A formula for calculating the peak runoff rate from a drainage area based on land use, soils, land slope, rainfall intensity, and drainage area.

Recession or receding limb (of a hydrograph) The portion of a hydrograph that occurs after the peak when the flow rate decreases.

Retention basin (wet pond) A reservoir, containing a permanent pool, for temporarily storing storm runoff and reducing the storm runoff rate from a drainage area.

Retaining wall A wall built to support a bank of earth.

Right of way The entire strip of land dedicated for highway purposes.

Riprap Stones or other material placed on a slope to prevent erosion by water.

Rising limb (of a hydrograph) The portion of a hydrograph preceding the peak when the flow rate increases.

Rough grade Stage of grading operation in which the desired landform is approximately attained.

Roughness coefficient *(n)* A factor in the Manning formula representing the effect of channel or conduit roughness on energy losses in the flowing water.

Runoff That part of precipitation carried off from the area on which it falls. Also, the rate of surface discharge of the above. (The ratio of runoff to precipitation is a coefficient, expressed as a decimal.)

Runoff curve number *(CN)* A parameter used in SCS hydrological techniques, based on soil characteristics and land use.

SCS Soil Conservation Service, a federal agency in the Department of Agriculture, dealing with erosion and flood control.

Sediment Solid material, both mineral and organic, in suspension, being transported, or having been moved from its original site, by air, water, gravity, or ice.

Sediment basin A reservoir formed by the construction of a barrier or dam built at a suitable location to permit the settling out of sediments (rock, sand, gravel, silt, or other material) before releasing the water.

Shallow concentrated flow Flow in shallow rills.

Shear stress (channel) Force per unit area exerted on the wetted area of a channel, acting in the direction of flow.

Sheet flow Flow over plane, sloped surfaces in a thin layer.

Shoulder The portion of roadway between the edge of the hardened wearing course and the ditch or embankment.

Sight distance The distance between approaching vehicles when first visible to one another on a horizontal or vertical curve.

Slide Movement of soil on a slope resulting in a reduced angle of repose usually occurring as a result of rainfall, high water, or thaw.

Slope The face of an embankment or cut section; any ground whose surface makes an angle with the horizontal plane.

Splash block A masonry block with its top close to the ground surface, which receives roof drainage and prevents erosion below the spout.

Storage (runoff) Runoff that is being temporarily impounded to permit controlling the runoff rate and/or improving water quality.

Storm sewer A conduit used for conveyance of rain water.

Structure Anything constructed that requires a permanent location on the ground or is attached to something that has a permanent location on the ground.

Subdrain A pervious backfilled trench containing a pipe with perforations or open joints for the purpose of intercepting groundwater or seepage.

Superelevation The rise of the outer edge of the pavement relative to the inner edge at a curve in the highway, expressed in feet per foot, intended to overcome the tendency of speeding vehicles to overturn when rounding a curve.

Swale A constructed or natural vegetated waterway.

Tangent A straight road segment connecting two curves.

Terrace An essentially level and defined area, usually raised, either paved or planted, forming part of a garden or building setting.

Time of concentration (T_c) The time for water to flow from the hydraulically most remote point in a drainage area to the point of interest.

Time of recession, T_{rec} The period of time from the peak of a hydrograph until it again reaches the beginning flow rate.

Time of rise (of a hydrograph), T_{rise} The period of time from the beginning flow rate until the peak flow rate is reached.

Travel time (T_t) The time for runoff to flow from one point in a drainage area to another.

Trench drain A linear structure that collects runoff from a paved area.

Water quality basin A reservoir which has a provision for removing pollutants from storm runoff by retaining the runoff from high-frequency storms (i.e., those with 1- or 2-yr frequencies) for prolonged periods (i.e., from 18 to 36 hr).

Watershed Region or area contributing to the supply of a stream or lake. (Also drainage basin or catchment area.)

Water table The level below which the ground is saturated.

Waterway A natural course, or a constructed channel, for the flow of water.

Weephole A small hole, as in a retaining wall, to drain water to the outside.

Weir An opening in the crest of a dam or embankment to discharge excess water; also used for measuring the rate of discharge.

Wetted perimeter The length of the wetted contact between the water and the containing conduit, measured along a plane that is perpendicular to the conduit.

Bibliography

American Concrete Pipe Association. 1980. Concrete Pipe Design Manual. American Concrete Pipe Association, Vienna, VA

American Iron and Steel Institute. 1980. Modern Sewer Design. American Iron and Steel Institute, Washington, DC

American Society of Civil Engineers, National Association of Home Builders, and the Urban Land Institute. 1990. Residential Streets, Second Edition. American Society of Civil Engineers, New York, National Association of Home Builders, Washington, DC, and ULI—the Urban Land Institute, Washington, DC

Asphalt Institute. 1984. Drainage of Asphalt Pavement Structures. The Asphalt Institute, College Park, MD

Barfield, B. J., Warner, R. C., and Haan, C.T. 1981. Applied Hydrology and Sedimentology for Disturbed Areas. Oklahoma Technical Press, Stillwater, OK

Cahill, T. H., Adams, M. C., and Horner, W. R. 1988. The Use of Porous Paving for Groundwater Recharge in Stormwater Management Systems. 1988. Floodplain/Stormwater Management Symposium, The Pennsylvania Department of Environmental Resources, State College, PA

Chow, Ven T. (Editor). 1964. Handbook of Applied Hydrology. McGraw-Hill Book Company, Inc., New York

Day, G. E., and Crafton, C. S. 1978. Site and Community Design Guidelines for Stormwater Management. College of Agriculture and Urban Studies, Virginia Polytechnic Institute and State University, Blacksburg, VA

Ferguson, B., and Debo, T. N. 1990. On-Site Stormwater Management: Applications for Landscape and Engineering, Second Edition. Van Nostrand Reinhold, New York

Goldman, S. J., Jackson, K., and Bursztynsky, T. A. 1986. Erosion and Sediment Control Handbook. McGraw-Hill Book Company, Inc., New York

Gray, D. H., and Leiser, A. T. 1982. Biotechnical Slope Protection and Erosion Control. Van Nostrand Reinhold Company, New York

Grzimek, G. 1972. Spiel und Sport in der Stadtlandschaft: Olympialandschaft Oberwiesenfeld 1972, Verlag Georg D. W. Callwey, Munich

Hughes, A. H. 1990. Making a Parking Lot into an Exhibit *in* The Public Garden, January 1990

Kassler, E. B. 1986. Modern Gardens and the Landscape, Revised Edition, The Museum of Modern Art, New York

Linsley Jr., Ray K., Kohler, M. A., and Paulhus, J. L. H. 1986. Hydrology for Engineers. McGraw-Hill Book Company, Inc., New York

Lynch, K. 1971. Site Planning, Second Edition. MIT Press, Cambridge, MA

Munson, A. E. 1974. Construction Design for Landscape Architects. McGraw-Hill Book Co., Inc., New York

Nathan, K. 1975. Basic Site Engineering for Landscape Designers. MSS Information Corp., New York

New Jersey Association of Conservation Districts. 1987. Standards for Soil Erosion and Sediment Control in New Jersey. New Jersey Department of Agriculture, Trenton, NJ

Parker, H., and Macguire, J. W. 1954. Simplified Site Engineering for Architects and Builders. John Wiley & Sons, Inc., New York

Pira, E. S., and Hubler, M. J. n.d. Tile Drainage Systems, University of Massachusetts, Amherst, MA

Poertner, H. G. (Editor). 1981. Urban Storm Water Management. Special Report No. 49. American Public Works Association, Chicago

Schroeder, W. L. 1975. Soils in Construction. John Wiley & Sons, Inc., New York

Schueler, T. R. 1987. Controlling Urban Runoff: A Practical Manual for Planning and Designing Urban BMPs. Metropolitan Washington Council of Governments, Washington, DC

Schwab, G. O., Frevert, R. K., Barnes, K. K., and Edminster, T. W. 1971. Elementary Soil and Water Engineering. John Wiley & Sons, Inc., New York

Schwab, G. O., Frevert, R. K., Edminster, T. W., and Barnes, K. K. 1981. Soil and Water Conservation Engineering. John Wiley & Sons, Inc., New York

Seelye, E. E. 1968. Data Book for Civil Engineers: Design, Third Edition. John Wiley & Sons, Inc., New York

Snow, B. 1959. The Highway and the Landscape. Rutgers University Press, New Brunswick, NJ

Stahre, P., and Urbonas, B. 1990. Stormwater Detention for Drainage, Water Quality, and CSO Management. Prentice-Hall Inc., Englewood Cliffs, NJ

Steele, F. 1981. The Sense of Place. CBI Publishing Co., Inc., Boston, MA

Sue, J. 1985. Landscape Grading Design, in Handbook of Landscape Architecture Construction, Second Edition, Volume I, M. Nelischer (editor). Landscape Architecture Foundation, Washington, DC

Untermann, R. K. 1978. Principles and Practices of Grading, Drainage and Road Alignment: An Ecologic Approach. Reston Publishing Co., Inc., Reston, VA

USDA—Agricultural Research Service. 1987. Agriculture Handbook No. 667, Stability Design of Grass-lined Open Channels. Washington, D.C.

USDA—Soil Conservation Service. 1986. Engineering Field Manual for Conservation Practices. National Technical Information Service, Springfield, VA

USDA—Soil Conservation Service. 1985. National Engineering Handbook Section 4: Hydrology. National Technical Information Service, Springfield, VA

USDA—Soil Conservation Service. 1986. Urban Hydrology for Small Watersheds (Technical Release Number 55). National Technical Information Service, Springfield, VA

USDOT—Federal Highway Administration. 1988. Hydraulic Engineering Circular No. 15, Design of Roadside Channels with Flexible Linings. McLean, VA

Whipple, W., Grigg, N. S., Grizzard, T., Randall, C. W., Shubinski, R. P., and Tucker, L. S. 1983. Stormwater Management in Urbanizing Areas. Prentice-Hall, Inc., Englewood Cliffs, NJ

Young, D., and Leslie, D. 1974. Grading Design Approach. Landscape Architecture Construction Series. Landscape Architecture Foundation, Washington, DC

Zolomij, R. 1988. Vehicular Circulation, in Handbook of Landscape Architecture Construction, First Edition, Volume II, M. Nelischer (editor). Landscape Architecture Foundation, Washington, DC

Index

Alignment:
    horizontal, 175
    vertical, 189
Andropogon Associates, 199, 202
Angle:
    deflection, 176, 177
    included, 176, 177
Antecedent precipitation factor, 108
Area drain, 133, 135
Average end area method, 162, 163

Backfilling, 161
Base, 159
Basin:
    catch, 133, 134
    detention and retention, 135
    dry, 90–93
    infiltration, 135
    sediment, 135
    water quality, 95
Basins, safety of, 93
Bayer, Herbert, 67
Bearing capacity soil, 37
Behnish, Günther, 73
Best management practices, 88
Borrow, 159
Borrow pit method, 165–167
Bulk excavation, 160

Catch basin, 133, 134
Center of curve, 176
Channel flow time, 102, 105
Channel flow, Manning's equation, 122
Cheek wall, 50
Circular curve:
    elements of, 176
    formulas, 176
Compaction, 159
Compound curve, 175, 176
Concentration, time of, 102, 104, 121
Constraints:
    critical, 43
    environmental, 35–37
    functional, 35, 38–43
Continuity equation, 140
Contour, 1
    area method, 163–165
    closed, 1, 6
    interval, 2
    line, 1
    lines, visualizing topography from, 29
    signature, 3
Contours and form, 1, 2
Critical flow depth, 143
Critical gradients, 42
Critical velocity, 143–145
Cross-pitch, 25
Crown:
    height, 25
    parabolic section, 26
    reverse, 26
    road, 25
    tangential section, 26
Culvert, 133, 134
Curb, 26
Curb:
    and gutter, 27
    batter-faced, 26
    beveled, 26
    mountable, 26
    rounded, 26
Curvature, point of, 176
Curve:
    broken-back, 175, 176
    center of, 176
    chord of, 176
    circular, 175, 176
    compound, 175, 176
    degree of, 178, 179
    high point of vertical, 192
    length of, 176
    low point of vertical, 192
    peak, 189
    radius of, 176
    reverse, 175, 176
    runoff, 123, 125–126
    sag, 189
    simple, 175, 176

Index

Curve (cont.)
    spiral, 176
    vertical, 189
Cut, 20, 159, 160
Cut and fill volumes:
    adjusting, 167, 168
    balancing, 168–171
    computation of, 162–171
Cut volumes, computation of, 162–171
Cut-off drain, 153

Deflection angel, 176, 177
Degree of curve, 178, 179
    arc definition, 178
    chord definition, 178
Depression, 4
Design storm, 93, 102
Design:
    and perception, 46–48
    architectonic, 45, 46
    geomorphic, 45, 46
    naturalistic, 45, 46
Detention:
    basin, 135
    facilities, 90–93
    storage volume, 128–132
Distance, runoff, 185, 186
Drain:
    area, 133, 135
    cut-off, 153
    inlet, 133, 134
    lateral, 51, 52
    trench, 133, 135
Drainage:
    area, 101, 102, 154
    coefficient, 153
    positive, 18, 43
    subsurface, 152–155
Drainage system:
    closed, 136
    combination, 136, 137
    open, 136
Drainage systems, design and layout, 137
Dry basin, 90–93

Earthworks Park, 67–70
Elevation:
    invert, 150
    spot, 10, 12, 61
Enclosure, 47
Environmental constraints, 35–36
Erosion and sedimentation, control of, 96–99
Excavation, bulk, 160

Fill, 20, 159, 160
Fill volumes, computation of, 162–171
Filter strip, 96
Fine grading, 161

Finished grade, 159
Flow:
    critical depth of, 143
    open channel, 122
    shallow concentrated, 121, 122
    sheet, 121
Foundation:
    continuous wall, 38, 39
    pole, 38, 39
    slab, 38, 39
    systems, 39
Functional constraints, 38–41
Gasworks Park, 70–72
Grade, 13
Grade change devices, 48
Grade, finished, 159
Gradient, 13
Grading design:
    architectonic, 45, 46
    geomorphic, 45, 46
    naturalistic, 45, 46
    and perception, 46–48
Grading plan, 60–62
Grading process, 53–59
    design development, 54
    design implementation, 54
    inventory and analysis, 54
Grading:
    and design, 67–83
    and enclosure, 47
    and landform design, 67–83
    athletic facilities, 41, 42
    by proportion, 28
    construction sequence for, 160, 161
    critical gradients, 42
    design, 45–48, 199–203
    fine, 161
    operations, 161, 162
    roads, 25
    sequence, 161
    standards, 42
Graphical peak discharge method, 123, 126–27
Graphics, grading plan, 60, 61
Grzimek, Günther, 73
Gutter, 26

Haag, Richard, 70
Horizontal alignment:
    construction drawing graphics, 182, 183
    for roads, 175–186
    procedure, 183–185
Horizontal sight distance, 182, 183
Hydraulic radius, 122, 123
Hydrographs, 108–114
Hydrologic:
    changes, 86, 87
    condition, 123
    cycle, 85, 86
    soil group, 123

Included angle, 176, 177
Infiltration:
    basin, 93, 94
    facilities, 93–95
    trench, 93, 95
Intensity, rainfall, 102, 103
Interpolation, 9
    between contour lines, 13
    between spot elevations, 10
    graphic, 11
Intersection, point of, 176
Invert elevation, 150

Kluska, Peter, 79

Length of curve, 176

Manhole, 135
Manning's equation:
    for channel flow, 122
    for sheet flow, 121
Modified Rational method, 108–112
Morris Arboretum, 199–203

Olympic Park, 73–78
Otto, Frei, 73
Overland flow time, 102, 104

Pavement, porous, 93–95, 96, 202, 203
Peak curve, 189
Peak discharge method, graphical, 123, 126–127
Perception, and slope, 46
Permeability, soil, 37
Permissible velocities, 139
Pipe flow, 149
Pipe systems, design and sizing of, 145–152
Point:
    of curvature, 176
    of intersection, 176
    of no cut, 22
    of no fill, 22
    of tangency, 176
Pond:
    retention, 88–90
    wet, 88–90
Porous pavement, 93–95, 96, 202, 203
Precipitation, antecedent, 108
Preparation, site, 160
Profile, 150
    road, 194, 195

Radius:
    hydraulic, 122, 123
    of curve, 176
Rainfall intensity, 102, 103

Ramp, 50, 51
Ramp, stair, 51
Rational method, 101–108
Recharge bed, storm water, 202
Retaining wall, 51, 52
Retardance factors, 139, 140
Retention:
    basin, 135
    pond, 88–90
Ridge, 3
Road alignment:
    design, 199–203
    horizontal, 175–186
    procedure, 194–196
Road grading, 25
Road, crown, 25
Roughness coefficients:
    for pipes and channels, 122
    for sheet flow, 121
Runoff coefficient, 102, 106
Runoff curve, 123, 125–126
Runoff curve numbers:
    for other agricultural lands, 126
    for urban areas, 125
Runoff volume diagrams, 112
Runoff:
    distance, 185, 186
    maximum rate of, 109
    peak rate of, 109
    storm, 85
    subsurface, 85
    surface, 85
    volume, 112

Saddle, of a swale, 27, 28
Sag curve, 189
Section, 3
Sediment basin, 135
Shallow concentration flow, 121, 122
Shear strength, soil, 37
Sheet flow, 121
    Manning's equation, 121
    roughness coefficients for, 121
Sight distance:
    horizontal, 182, 183
    vertical, 194
Site preparation, 160
Slope, 13
    analysis, 17
    and perception, 46
    concave, 4, 46
    convex, 4, 46
    diagram, 54
    direction of, 6
    express in degrees, 15
    expressed as ratio, 15
    formula, 14
    rounding, 23
    uniform, 4

Slopes, as grade change device, 52, 53
Soil Conservation Service:
    24-hr rainfall distribution, 118
    hydrologic procedures, 117
    rainfall distribution patterns, 117, 118
    TR55, 117–132
Soil erosion factors, 96, 97
Soils:
    properties of, 37
    well-graded, 37
Spot elevation, 10, 12, 61
Stairs, 48–50
Stationing, 179, 181–182
Storm frequency, 102
Storm runoff, 85
Storm water management:
    landscaping practices, 95, 96
    philosophy, 87, 88
    principles and techniques, 88–96
    systems, 133–137
Storm water, recharge bed, 202
Storm:
    design, 93, 102
    frequency, 102
Street patterns, 39–41
Sub-base, 159
Subcritical velocity, 143
Subgrade, 159
    compacted, 159
    undisturbed, 159
Subsurface drainage, 152–155
    depth and spacing, 153, 154
    drainage area, 154
    drainage coefficient, 153
    patterns, 152, 153
    system design, 152–155
Subsurface runoff, 85
Summitt, 4
Supercritical velocity, 143
Superelevation, 185, 186
Surface drainage, 137
    collection, 137
    conduction, 137
    disposal, 137
Surface runoff, 85
Surfacing, 161
Swale, 26, 133
    depth of, 27
    design, 138–145
    high point of, 28
    saddle, 27
    sizing, 138–145

Tabular hydrograph method, 127, 128
Tangency, point of, 176
Tangent, 175
Tangent distance, 176
Terrace, 53
    bench, 53
    drainage, 53
    grading, 20
    on fill, 20
    sections, 21, 53
Time of concentration, 102, 104, 121
Time of recession, 109
Topographic map, 1
Topography and street patterns, 39–41
Topsoil, 159
Travel time, 121
Trench drain, 133, 135
Trench, infiltration, 93, 95

Underdrainage, 152

Valley, 3
Velocity:
    critical, 143–145
    permissible, 139
    subcritical, 143
    supercritical, 143
Vertical alignment:
    for road, 189–196
    construction drawing graphics, 193, 194
Vertical curve:
    elements of, 190
    equal tangent, 190, 191–192
    formula, 190
    unequal tangent, 190, 191
Vertical sight distance, 194
Volume:
    detention storage, 128–132
    storage for detention or retention ponds, 113, 114

Wall:
    cheek, 50
    retaining, 51, 52
Water quality basin, 95
Watershed, 101
Weephole, 51, 52
Westpark, 79–83
Wet pond, 88–90